AF625100

The International Library of Bioethics

Founding Editors

David N. Weisstub
Thomasine Kimbrough Kushner

Volume 114

Series Editor

Dennis R. Cooley, North Dakota State University, School of Humanities, Fargo, ND, USA

Editorial Board

Terry Carney, Faculty of Law Building, University of Sydney, Sydney, NSW, Australia

Marcus Düwell, Universiteit Utrecht, Utrecht, The Netherlands

Elizabeth Heitman, Southwestern Medical Center, University of Texas, Dallas, TX, USA

David Augustin Hodge, National Center for Bioethics, Tuskegee University, Tuskegee Institute, AL, USA

Søren Holm, Centre for Social Ethics and Policy, The University of Manchester, Manchester, UK

Nora L. Jones, Center for Urban Bioethics, Temple University, Philadelphia, PA, USA

Gerrit Kimsma, Radboud UMC, Nijmegen, The Netherlands

Daniel P. Sulmasy, M. D., Edmund D. Pellegrino Center for Clinical Bioethics, Washington, DC, USA

The *International Library of Bioethics* comprises international and interdisciplinary volumes on foundational and applied issues in bioethics. The series focuses on perennial topics of aging, mental health, preventive medicine, medical research issues, end of life, biolaw, and other areas of bioethics, whilst expanding into other current and future questions involving the living world. The *International Library of Bioethics* examines the role of bioethics against the background of increasing globalization and interdependency of the world's cultures and government.

The series explores meaningful ways to answer the questions of how to direct biotechnology to human and other living things' ends, how to deal with changed values in the areas of religion, society, and culture, and how to formulate a new way of thinking. That is, to create a new, practical bioethics.

The international Library of Bioethics welcomes book proposals representing the broad interest of the series' focus. We especially encourage proposals addressing aspects of changes in biological and medical research and clinical health care, health policy, medical and biotechnology, and other applied ethical areas involving living things, with an emphasis on those interventions and alterations that force us to re-examine foundational issues.

Bryan Pilkington

The Medical Act: Conscientious Practice in a World of Dissention and Disagreement

Springer

Bryan Pilkington
Department of Medical Sciences
Hackensack Meridian School of Medicine
Nutley, NJ, USA

ISSN 2662-9186 ISSN 2662-9194 (electronic)
The International Library of Bioethics
ISBN 978-3-032-00047-7 ISBN 978-3-032-00048-4 (eBook)
https://doi.org/10.1007/978-3-032-00048-4

© The Editor(s) (if applicable) and The Author(s), under exclusive license to Springer Nature Switzerland AG 2025

This work is subject to copyright. All rights are solely and exclusively licensed by the Publisher, whether the whole or part of the material is concerned, specifically the rights of translation, reprinting, reuse of illustrations, recitation, broadcasting, reproduction on microfilms or in any other physical way, and transmission or information storage and retrieval, electronic adaptation, computer software, or by similar or dissimilar methodology now known or hereafter developed.
The use of general descriptive names, registered names, trademarks, service marks, etc. in this publication does not imply, even in the absence of a specific statement, that such names are exempt from the relevant protective laws and regulations and therefore free for general use.
The publisher, the authors and the editors are safe to assume that the advice and information in this book are believed to be true and accurate at the date of publication. Neither the publisher nor the authors or the editors give a warranty, expressed or implied, with respect to the material contained herein or for any errors or omissions that may have been made. The publisher remains neutral with regard to jurisdictional claims in published maps and institutional affiliations.

This Springer imprint is published by the registered company Springer Nature Switzerland AG
The registered company address is: Gewerbestrasse 11, 6330 Cham, Switzerland

If disposing of this product, please recycle the paper.

For Jennifer Laura Sewell

sine qua non

Acknowledgements This book has its origins in a number of conversations over the last decade. I am grateful to colleagues, friends, and students for comments and engagement on a variety of topics addressed in the volume. Many thanks to Jason Eberl and Jeff Bishop for early encouragement and helpful critiques of my thinking in this space. Thanks to Gilbert Meilaender and to Lauris Kaldjian for a series of discussions on the topic and a special thanks to Dennis Cooley, Ryan Nash, Lynette Martins, and Jeff Byrnes for engaging questions and helpful commentary when parts of the book were presented. I am also grateful for the many conversations over many years with my father, Bryan G. Pilkington, about medicine and ethics. I have also learned much from Tom Harter and Paul Cummins on the subject.

Books are not written without time and encouragement, and I am grateful to have a department chair who understands the importance of scholarship and the time it takes to do it properly, so thank you to Stanley Terlecky, and all of my engaging Department of Medical Sciences colleagues.

A significant debt is owed to David Solomon, who offered helpful comments and critiques on a number of chapters when they appeared as conference presentations, most especially Chapter 4. A similar debt is owed to Mark Cherry for arguments in Chapter 2, and to Charles Binkley for arguments in Chapter 5's section on AI.

Books are not written without others bearing additional costs. The greatest debt to be paid is to Jennifer, Vera, and Alessandra; Jennifer for her support, Vera for her patience, and Alessandra for her willingness to accompany a writer, even into the early morning hours.

Competing Interests The author has no competing interests to declare that are relevant to the content of this manuscript.

Preface

My interest in questions of conscience arose due to the interprofessional and multidisciplinary spaces I have found myself in over the last decade. Three stories, which roughly track the narrative arch of my thought on the subject are illustrative and so I share them in this preface.

The first story comes from a time early in my teaching career and involves an ill-fitting request. I was a somewhat newly minted PhD and eagerly anticipating teaching my first courses outside of the institution where I did my graduate work. Fortunately—though it felt unfortunate then—funding was predicated upon teaching as a graduate student, so I ended up teaching a lot of courses in the years I was in school. The experience and lecture notes ended up being quite useful, as well as the opportunities to serve as a teaching assistant and learn a variety of teaching styles, pedagogical influences, and content emphases (even for similarly titled courses). And so, it was with a fairly decent amount of teaching experiences and perhaps a bit more confidence than was anticipated by those who hired me that I arrived to teach a fairly wide set of philosophy courses. What my graduate school training did not prepare me for was a request for a practice which I had not seen before: use "scantron" tests to evaluate student performance in humanities-based courses. This request accompanied the notice that one of the classes I was asked to teach would be a course with a significantly larger number of students than was common for the course. This number was larger than the cap for similar courses at other institutions and well beyond numbers with which I was familiar teaching. I baulked at the request. Bracketing student experience and faculty workload considerations, what really concerned me was how to pull off teaching a large philosophy course, from a logistical perspective. How does one put that many students in a situation to succeed in a course where they could not all have sufficient time to speak and debate, to engage their classmates' ideas, to try out arguments, to revise or confirm their own untutored assumptions? Assessment was almost an afterthought, but it became the focus of pre-course conversations with the administration: How could such a large number of students be meaningfully assessed in this way?

The response to my queries was simple: offer lectures and assess student performance via scantron tests. Though such examinations probably have other names, these tests were standardized tests wherein students could fill in bubbles to indicate an answer to a multiple-choice question and a machine, which the institution possessed, would read the bubbles and grade the exams. Oddly enough for a project on conscientious objection (CO) in medicine, this is its origin. I simply could not lecture at students, have a scantron test evaluate them, and then say they had taken a proper course in philosophy. There may very well be disciplines, which possess content that can be absorbed in this way and an apprehension thereof which can be evaluated by filling in bubbles, but philosophy is not one of them. Why this led to an interest in CO in medicine is that I objected to teaching the course in this way and did so for reasons that I rooted in a disciplinary-specific expertise. Practitioners of particular disciplines or members of particular professions, in a conversation with other members engaged in those practices,[1] ought to (at least to a very large degree) set the bounds of the discipline and its methods. Setting such bounds includes how its content (in a broad sense) is explained to others and passed on to future practitioners. Thus, as a philosopher, I thought I should be able—with advice from and in conversation with other philosophers—say what counts as relevant to philosophical pursuits or, at least, help to shape the bounds of particular features of the practice of philosophy. I retreat to this weaker claim because, of course, philosophical pursuits are, in many ways, human pursuits, and as confident as a new PhD might be I was not *that* confident. Yet I was confident enough to reject the idea that a philosophical discourse around challenging and perennial topics could be evaluated by a scantron test. Courses in philosophy—whatever the topic—do not suit students well if students do not have opportunities to write essays, discuss ideas with their classmates, debate and "try on" new ways of thinking.

My argument was eventually successful, which is to say I was able to teach one too many sections of that course with a large but average number of students. However, that initial conversation still lingered. My objections to the request were not that the requestor did not have the relevant institutional authority to demand such a thing nor that they did not do so in the right way, per the policies of the institution, but that practicing professionals who are recognized as experts in particular fields ought not to be required to violate the best practices of those fields. One cannot "do" philosophy, which includes teaching philosophy, in the manner that was suggested. One response to my claims is important to consider: it might simply be that there are different ways of doing philosophy. This is surely true, but though the diversity of philosophical methods and practices may be wide, they cannot be reduced to mere memorization of terms and concepts. To miss the argumentative nature of philosophical discourse is to miss out on something that is necessary. The objector might continue to push, suggesting that I simply hold an outdated view of the subject: there might have been necessary and sufficient conditions for some activity to be properly argumentative as to count as philosophical was a fair description of what teaching a course in philosophy was; however, that it is no longer the case. A paradigm shift occurred and now teaching philosophy is something else. The student (or some other payer),

[1] For those familiar with the work of the philosopher Alasdair MacIntyre, I mean practices in that technical sense. See MacIntyre (2007).

who should be understood as a consumer, might order a different kind of course and it ought to be provided. I did not think that was right then, nor do I think it is right now,[2] and this line of thinking applies both to teaching philosophy and to practicing medicine. Reasonable persons can disagree, and this is not a book on how to best teach a course in philosophy. I hope that knowing where the author stands on such matters will shed some light on the arguments in the book and place the reader in a better situation to evaluate the arguments in the following chapters. With that in mind, I share that I subscribe to a view, one which the political theorist Michael Rosen[3] articulates well in his book on dignity. To paraphrase Rosen's description in the introduction to that text: philosophy without arguments is like football without the ball. Thus, one could not evaluate someone's understanding of philosophy without a discussion, without the person demonstrating some engagement with arguments. To do so would be like reviewing the relevant statistics before a match: the forward is 6'4 with a significant speed advantage, but the goalie has a reach that… that is why, as one famous sports commentator has noted, they play the game.

So, it was with an early defense of a space for (at least some) practitioners to exercise their best judgment as authoritative that I fell into thinking more and more about considerations of conscience.

A second point in the journey of this project and in my thinking about conscience is due to my work in interprofessional practice. Though my training is in philosophy, I have spent a good deal of time with professionals from other disciplines. First, I spent time with those who populated other liberal arts departments—finding similarities in conversations with the political theorists and the biologists and differences with the historians and the professors of business. Second, I spent time with practitioners from various health professions. This interprofessional work, which was more challenging in the second space, though similarly rewarding, arose through teaching and research. Learning how similar and how different teaching a practical ethics course could be in a medical or nursing context as opposed to a speech language pathology or occupational therapy context was surprising. Research with other professionals equally rewarding but not surprising: collaborative projects that took up questions germane to a variety of disciplines or required both the work of an ethicist and the experience of a clinician fit the anticipated models. It was through this teaching and research that I came to better understand the differences and similarities between distinct health professions, how they are practiced, and how individual professionals view themselves and what they are doing. This was apparent both in particular requests for "ethics" courses or sessions in a nursing program as opposed to an athletic training program, but also in the kinds of research that could be produced in each space and, more importantly, what was valued in those spaces.

---

[2] This is not to say that courses should not be significantly revised in future iterations of their delivery—philosophy is not static; nor is it to say that student input is not essential to that revision—it is.

[3] Rosen, Michael (2012).

One especially formative example comes from research on cardiac devices and ethical guidance about their deactivation. My work on this subject[4] was sparked by a request for assistance in a case that illustrated all of the elements that can lead to moral distress among clinicians. However, what has been most influential on my work (though I am sure not the influences that were most impactful for the clinician, patient, or their family in the case) was surprising to me. It was not the arguments about the status of pacemakers in patients who are fully dependent on them. It was not reflection on what the proper action description of a deactivation is in such cases as opposed to less dependent cases or cases involving related but distinct technology, such as defibrillators. What was most influential on my work in this area was the scholarship on perspectives of different parties to these conversations. It was that sociological work[5] and putting those perspectives into conversation with the perspectives of others—including consensus statements that are generated and promulgated by associations or societies of professionals[6]—that cemented my interest in considerations of conscience in healthcare and, eventually and particularly, in conscientious practice in medicine. Health professionals, all claiming to be part of a single care team and all truly being involved in the care of patients, employed not only different approaches (distinct to their disciplines and their own personalities) but they also deeply disagreed with each other about the particular features of that care and practices associated with it. Especially interesting for me was the fact that they disagreed, at least in some instances, because of underlying differences rooted in divergent ethical principles. For example, in cardiac device deactivation, a philosophical distinction between withholding treatment and withdrawing treatment appeared to influence the health professionals in both nursing and in industry (the "techs"), whereas this distinction was rejected by physicians.[7] Better understanding the perspectives of health professionals through working on the ethics of cardiac device deactivation and seeing where agreement existed and where professional opinions diverged was eye-opening. As someone who came to this work with an untutored (except via personal experience) view about the spaces in which practitioners should receive deference in light of their expertise, I was forced to rethink that view. How should professions and the individuals that comprise them be understood relative to what some would call professional autonomy (though I argue against autonomy as the basis for conscience practice in chapter four)? Does that autonomy inhere in individuals? Does it apply to professions, as a whole? If it applies to individuals, how are we to understand professions that seek a shared goal (the health of the patient), by jointly accepted means (the methods and resources of nursing, for instance), but in which individuals disagree about what is right? How do we practically move forward without constant pauses in care, workarounds, and complications in clarity and consistency? If professional autonomy only applies to

---

[4] Pilkington (2020).

[5] On this topic, the following were illustrative: Kramer et al. (2011), Mueller et al. (2008, 2011) and Pedersen et al. (2013).

[6] On this topic, the following was illustrative: Lampert et al. (2010).

[7] For more on these perspectives and how the categories were constructed, see Pilkington (2020).

professions (as a whole, as opposed to individuals), where are we to turn to find out what professions have to say on any of these subjects? Professional societies or associations are an obvious choice, but one which I argue against later in this text (see chapter two). I was left with more questions than answers and as I tried to wrestle with them, I realized that in portions of my interprofessional work, especially in the interprofessional research collaborations), I found myself to be an outsider. I do not hold a clinical degree; I was not trained in a health profession, I do not "have a practice" in addition to my teaching and research. I am, in these contexts, "a PhD ethicist" or that is how I have been introduced to clinician educators in these spaces, marking a distinction between us. This is not a negative distinction; it is just a difference. Not being in the club of the relevant health profession under consideration has afforded me a degree of freedom and, as a result, a certain amount of critical distance. There is surely much that I miss due to this orientation, but there are also things I see that those who are "in it" do not, or cannot, see. For me, this critical distance has been helpful in understanding disagreements that arise and in allowing me to bring some of the tools of philosophy to this arena. In addition to those tools, I maintain a strong interest in the philosophy of disagreement, in particular, how we live in communities with moral disagreement. This leads to the third anecdote.

The final story which serves as background for the work in this book comes from my interest in understanding how people live with moral disagreement. The subject is the name of a course I co-taught a few years ago with friend and colleague, Molly Patterson, who studies political theory. In many ways this is the part of the story that most directly relates to this work on conscience in medicine. Michael Rosen, the aforementioned political theorist whose introduction to his book on dignity so nicely framed the importance of arguments to philosophical pursuits, illustrates why philosophical training can be useful but also challenging in helping one to understand how best to live with others when deep disagreements exist. A second metaphor in that introduction compares philosophy to a game of chess,[8] that is, one must understand all of the rules and anticipate moves any number of steps down the line in order to be successful. That is where the similarities end, however, because, as Rosen notes, in philosophy—as opposed to chess—opponents get to fight about the rules themselves. Chess is challenging, but also fun and rewarding. Philosophy requires additional groundwork. Arguing whether the knight ought to move up three and over one, instead of up two and over one, before one commences with the game is not an approach to life that is for everyone; however, it does force those playing to engage other perspectives in a deep way and to take others seriously.

I mention this third story for two reasons. First, it illustrates for the reader yet another set of my biases: I find philosophical approaches to ethical challenges to be terribly useful. This means that there are costs I am willing to bear, such as additional time to think through challenges, conversations with others to dig deep and get at what differences might exist, and an acceptance of the tension that can arise in teams or on collaborative projects when arguments are to be taken seriously as a matter of course. In the fast-paced world of healthcare practice and research, this time is not

[8] Rosen (2012).

always available. Second, it helps to explain why I have chosen some interlocutors for the conversations in this book and not others or, at least, why some arguments and positions within the literature on conscience will have appeared so interesting (and in need of critical engagement) to me. One such position was articulated in a recent *New England Journal of Medicine* article by Ronit Stahl and Ezekiel Emanuel. They argue against allowing space for conscientious practice within healthcare, relying on philosophical resources which should point in a different direction.

With that background in place, let us turn to the book itself. The aim of this text is to examine conscientious practices in healthcare. Discussions of conscience in healthcare are often framed as involving situations where professionals object to legally available services and where there is social allowance of some subset of such objections due to a respect for professionals' beliefs. This framing has led to a good deal of work on the subject with the aim of adjudicating disagreements between allowing persons to practice according to their own beliefs and the costs of such allowances. Responses to this challenge range from significant restrictions on a professional's spheres of liberty to (almost) complete free reign from professionals, with a plethora of compromise positions, variously nuanced, in between. Robert Card, for instance, focuses on reasonability in his articulation of the issues and his attempt to resolve them.[9] Mark Aulisio and Kavita Arora rely on autonomy and reciprocity to parse out what they see as different claims of conscience as a means to resolving tensions—both theoretical, as well as practical.[10] Some, such as Stahl and Emanuel[11] and Julian Savulescu[12] argue that religious views are central to the issue and deeply problematic, while others, such as Mark Cherry find the rejection of religious views to itself be problematic.[13] David Hershenov addresses issues relevant to this discussion through the idea that medicine possesses an internal morality,[14] whereas Lauris Kaldjian[15] takes pains to articulate the role of practical wisdom for clinicians in their relationships with patients. Daniel Sulmasy[16] focuses on what conscience is, as opposed to beginning with the issue created by respecting what many call CO in healthcare. Mark Wicclair,[17] perhaps most famously and influentially, offers a view that places integrity at the center of debates over conscience in healthcare.

This text takes a different approach. It seeks to take a step back to consider whether conscientious practice—the reliance on a professional's best judgment—can be defended. Framing the relevant issues and questions in terms of conscientious practice allows for a new look at an issue that has stalled of late. In particular,

[9] Card (2020).

[10] Aulisio and Arora (2014).

[11] Stahl and Emanuel (2017).

[12] Savulescu (2006).

[13] Cherry (2012).

[14] Hershenov (2021).

[15] Kaldjian (2014).

[16] Sulmasy (2008).

[17] Wicclair (2011).

it allows for the rethinking of the status of religious claims within healthcare, the role of disagreement within professions, and ultimately, asks the question—which is decisive in this conversation—what is a medical act?

In this book, I argue that the entrenched disagreement within the debate over CO is due, in part, to an understandable but mistaken framing of the relevant issues. Most consider that cases in which healthcare practitioners find themselves to be at odds with a judgment within medicine for non-clinical reasons to be cases where they must follow the professionally supported standard of care and patient request or where they might be allowed to conscientiously object. This framing has led to descriptions of this kind of disagreement as dependent upon personal views of medical professionals being foisted on patients. This dialectic allows for creative work to be done to match professionals with patients, to exempt some practitioners from offering care in some cases, and to restrict some patient aims in light of available (human) resources. The creative work is limited to these realms because this mistaken framing has led to deeply entrenched, oppositional positions.

In addressing whether the conscientious practice of individual medical professionals is coherent and acceptable on its own, I take a few steps. The first step, the primary task occupying this book, is to describe the mistaken framing and to articulate a better one. In order to do this, I examine the debate (Chapter 1) and frame discussions in light of conscientious practice and conscience dissension. With this framing, I move to discuss disagreement within healthcare professions, focusing primarily on the profession of medicine (Chapter 2). That discussion helps to uncover a key confusion in much of the debate, which centers on religion. Though many claim that religious beliefs give rise to the situations of CO, I argue that differing accounts of moral responsibility are at the root of this disagreement (Chapter 3). I take up the topic of moral responsibility in Chapter 4, situating this idea vis-à-vis arguments in the space that prize autonomy over other defining ethical principles. The work in Chapter 4 leads to a discussion of what is going on when judgements are made in healthcare: the medical act (Chapter 5). Chapter 5 also houses work on a new and extremely interesting topic connected with conscience: artificial intelligence (AI).

If the arguments in this book are successful, there will be new avenues by which to move forward, in light of disagreement, and also reason for hope for better and more cohesive healthcare responses to patients in need.

Nutley, NJ, USA Bryan Pilkington

References

Aulisio, Mark P., and Kavita Shah Arora. 2014. Speak No Evil? Conscience and the Duty to Inform, Refer or Transfer Care. *HEC Forum: An Interdisciplinary Journal on Hospitals' Ethical and Legal Issues* 26 (3): 257–266. https://doi.org/10.1007/s10730-014-9242-8.

Card, Robert F. 2020. *A New Theory of Conscientious Objection in Medicine: Justification and Reasonability*. Routledge Annals of Bioethics 21. New York: Routledge.

Cherry, M. J. 2012. Conscience Clauses, the Refusal to Treat, and Civil Disobedience-Practicing Medicine as a Christian in a Hostile Secular Moral Space. *Christian Bioethics* 18 (1): 1–14. https://doi.org/10.1093/cb/cbs009.

Hershenov, David. 2021. Conscientious Objection or an Internal Morality of Medicine? *Christian Bioethics: Non-Ecumenical Studies in Medical Morality* 27 (1): 104–121. https://doi.org/10.1093/cb/cbaa020.

Kaldjian, Lauris Christopher. 2014. *Practicing Medicine and Ethics: Integrating Wisdom, Conscience, and Goals of Care.* New York: Cambridge University Press.

Kramer, Daniel B., Abigale L. Ottenberg, Samantha Gerhardson, Luke A. Mueller, Sharon R. Kaufman, Barbara A. Koenig, and Paul S. Mueller. 2011. 'Just Because We Can Doesn't Mean We Should': Views of Nurses on Deactivation of Pacemakers and Implantable Cardioverter-Defibrillators. *Journal of Interventional Cardiac Electrophysiology: An International Journal of Arrhythmias and Pacing* 32 (3): 243–252. https://doi.org/10.1007/s10840-011-9596-7.

Lampert, Rachel, David L. Hayes, George J. Annas, Margaret A. Farley, Nathan E. Goldstein, Robert M. Hamilton, G. Neal Kay, et al. 2010. HRS Expert Consensus Statement on the Management of Cardiovascular Implantable Electronic Devices (CIEDs) in Patients Nearing End of Life or Requesting Withdrawal of Therapy. *Heart Rhythm* 7 (7): 1008–1026. https://doi.org/10.1016/j.hrthm.2010.04.033.

MacIntyre, Alasdair C. 2007. *After Virtue: A Study in Moral Theory*, 3rd ed. Notre Dame, Ind: University of Notre Dame Press.

Mueller, Paul S., Sarah M. Jenkins, Katrina A. Bramstedt, and David L. Hayes. 2008. Deactivating Implanted Cardiac Devices in Terminally Ill Patients: Practices and Attitudes. *Pacing and Clinical Electrophysiology: PACE* 31 (5): 560–568. https://doi.org/10.1111/j.1540-8159.2008.01041.x.

Mueller, Paul S., Abigale L. Ottenberg, David L. Hayes, and Barbara A. Koenig. 2011. 'I Felt like the Angel of Death': Role Conflicts and Moral Distress among Allied Professionals Employed by the US Cardiovascular Implantable Electronic Device Industry. *Journal of Interventional Cardiac Electrophysiology: An International Journal of Arrhythmias and Pacing* 32 (3): 253–261. https://doi.org/10.1007/s10840-011-9607-8.

Pedersen, Susanne S., Rismy Chaitsing, Tamas Szili-Torok, Luc Jordaens, and Dominic A. M. J. Theuns. 2013. Patients' Perspective on Deactivation of the Implantable Cardioverter-Defibrillator near the End of Life. *The American Journal of Cardiology* 111 (10): 1443–1447. https://doi.org/10.1016/j.amjcard.2013.01.296.

Pilkington, Bryan C. 2020. Treating or Killing? The Divergent Moral Implications of Cardiac Device Deactivation. *The Journal of Medicine and Philosophy: A Forum for Bioethics and Philosophy of Medicine* 45 (1): 28–41. https://doi.org/10.1093/jmp/jhz031.

Pilkington, Bryan C. 2021. Erratum to: Treating or Killing? The Divergent Moral Implications of Cardiac Device Deactivation. *The Journal of Medicine and Philosophy: A Forum for Bioethics and Philosophy of Medicine* 46 (3): 377–377. https://doi.org/10.1093/jmp/jhaa001.

Rosen, Michael. 2012. *Dignity: Its History and Meaning*. Cambridge, Mass: Harvard University Press.

Savulescu, Julian. 2006. Conscientious Objection in Medicine. *BMJ* 332 (7536): 294–297. https://doi.org/10.1136/bmj.332.7536.294.

Stahl, Ronit Y., and Ezekiel J. Emanuel. 2017. Physicians, Not Conscripts — Conscientious Objection in Health Care. *New England Journal of Medicine* 376 (14): 1380–1385. https://doi.org/10.1056/NEJMsb1612472.

Sulmasy, Daniel P. 2008. What Is Conscience and Why Is Respect for It so Important? *Theoretical Medicine and Bioethics* 29 (3): 135–149. https://doi.org/10.1007/s11017-008-9072-2.

Wicclair, Mark R. 2011. *Conscientious Objection in Health Care: An Ethical Analysis*. Cambridge: Cambridge University Press.

Contents

Chapter 1
Conscience Dissenters and Conscientious Practice

What is the proper role of conscience in medicine? This is the question that this book takes up, however it does so in a less familiar way. This question is asked with significant concern, and increasing so, given the social divisiveness that continues across the globe. However, the commonly presupposed categories into which answers to this question must fall, are not accepted. Asking about the proper role of conscience in medicine is another way of asking whether CO should be allowed. Though the arguments in this book support something like this idea, it is mistaken to think that the kinds of practices that those arguments engage are all properly understood as instances of CO. The familiar reading of the question also suggests that there is a thing (i.e., medicine) and that the role of a professional's conscience may either fit or not. The familiar interpretation of this commonly asked question captures much of the oft-expressed sentiments, starting places, and assumptions of those arguing in this space; however, it fails to capture the integrated nature of a person's conscience with what they do. The question suggests that one might answer negatively: there is no place for conscience in medicine. However, persons cannot turn their consciences on and off, and so this raises concerns about the bifurcation of identity when one is a professional. Though that on/off switch imagery may be used to explain lapses in moral judgment (or failures of conscience), it is just that, imagery. This may explain, in part, why proponents of some positions, best characterized as conscience dissenters (CDs), argue against allowing space for conscience in medicine and couple these arguments with the claim that if one is not willing to restrict one's conscience in certain ways (turn it off), then one ought to leave the profession of medicine. It is not easy to maintain standards of a profession, take seriously professional identity and professional identity formation, and at the same time ask professionals to ignore what these other pursuits are said to nurture.

The question is also suggestive of positive answers: the conscience of professionals ought to have a role in medicine. However, this too suggests that one might have it or not have it, allow practitioners to use it or not to use it, and the like. The issue with asking the question in this way is that it leads to thinking about the

© The Author(s), under exclusive license to Springer Nature Switzerland AG 2025

B. Pilkington, *The Medical Act: Conscientious Practice in a World of Dissention and Disagreement*, The International Library of Bioethics 114,
https://doi.org/10.1007/978-3-032-00048-4_1

conscientious practice of individual healthcare professionals as something fitting or ill-fitting to a situation: either one practices according to certain norms or one is allowed to appeal to one's conscience in a small subset of instances in which one might reject those norms yet wish to remain a part of a community. Though practical policies might proscribe all but these options, they miss out on an important feature of conscience: the scope of its application. If we are to defer to persons in some instances because their consciences are troubled by their performance of particular actions or by their connection with the performance of particular actions, how do we decide to what kinds of things claims of conscience can legitimately be applied? Notice, the question here is not asking about when is a claim of conscience morally permissible, legally acceptable, or practically feasible. Rather, it is asking what properly falls within the domain of conscience. The former interpretation is about success or failure: when will a claim of conscience be heard and respected? Answers to this question so interpreted would delineate the necessary and sufficient conditions for moral permissibility, legality, or practicality. The latter interpretation is about proper standing. Answers to this question so interpreted would address something closer to legitimacy. Two important points follow from this discussion: the first is about moral responsibility, the second is about integrity. Each is taken up, in turn.

It is a primary contention of this book that many discussions of conscience are misguided. It is the hope of this book that by refocusing these discussions, some of the entrenchment within the debate might be avoided, allowing space for new resolutions. One example of the misguided nature of this debate is related to how persons and professions are understood, how actions (and omissions) are thought of, and what the relevant sources of conscience might be. Consider the claim that a medical professional[1] ought to check their religious views at the patient's door. This adage, or something like it, has been expressed by a variety of CDs in various ways.[2] The things that motivate expressions like this are reasonable. What is unreasonable, or what is in need of further explication and argument, is the assumption that the relevant persons, their professions, their beliefs and practices, and the sources of those beliefs ought to fit cleanly into the description of a request to practice conscientiously. The position articulated in this text is nonstandard. The standard description of cases fitting the adage are of the form: *Dr. Knowles is uncomfortable offering medical aid in dying (MAID) to patients, wherein Dr. Knowles is devoutly religious and religious reasons are what motivate her to not participate in MAID, which she understands to be ending a patient's life, a life which is a gift from God.* The standard conversations ensue. Questions about the sincerity of Dr. Knowles' beliefs abound: does she genuinely believe *x*? And is this belief held deeply? Religious practices ought to be respected, so deference should be given to Dr. Knowles; however, medical practice is not the appropriate place for deference to religion…is it? Should not Dr. Knowles

[1] Many, if not all, of the arguments in this book apply to members of any well established health profession. However, both for ease of phrasing and for clarity of examples and lines of argument, *physicians* and *medicine* are used throughout.

[2] Prominent examples include: Stahl et al. (2017) and Savulescu (2006).

keep her religion to herself? Her exercise of those beliefs will be respected, as long as they remain outside of this professional context and do not impact others…

The case of Dr. Knowles is a useful one, both to remind the reader of the common kinds of cases in this space, but also as a way of highlighting some of the differences between more standard ways of thinking about conscience in medicine and the account offered in this text. In the case of Dr. Knowles, the CDs (a cluster of positions that oppose or seek to restrict the role of conscience in medical practice) object, and they do so for a number of reasons. Some argue that religion or religious practice has no place in medicine.[3] Others will argue that if Dr. Knowles wishes for her religious beliefs to be respected, then she ought to respect the beliefs of others and confine her beliefs to other spaces or ensure that they do not impact others.[4] Some are primarily concerned with the foisting of beliefs onto patients, others with the deviation from profession-specific norms, and still others with the resultant variations in care offerings (making it challenging for vulnerable persons to easily access the care they need).

Contrary to CDs who give priority to other normative features (for example, laws, profession-specific norms, or the claims of others), conscience protectors (CPs) give primacy of place to the conscientious practice of individual health professionals. They may argue for this primacy outright or offer a compromise which carves out some space for the exercise of conscience by professionals. CPs are often concerned about the integrity of persons. This focus within the CO literature, more broadly, is due to the work of many contributors, but most especially Mark Wicclair.[5] The motivating idea for advocates of integrity-based conscience protection is that the beliefs of persons that are tied to their self-identity ought to be respected, at least as far as is socially and practically feasible. For Dr. Knowles, for instance, to medically aid someone in dying might be morally wrong. Additionally, for her, participation in the practice might be wrong enough that it would harm her own understanding of who she is. It is this latter point that it at the crux of integrity-based CP. Somewhat along these lines, Dan Sulmasy has described conscience "as having two interrelated parts: (1) a commitment to morality itself; to acting and choosing morally according to the best of one's ability, and (2) the activity of judging that an act one has done or about which one is deliberating would violate that commitment."[6] Sulmasy also notes, "Conscientious living is not easy. Following conscience means following through on this fundamental commitment to be moral. To lack this fundamental commitment to moral integrity is to be less than fully human."[7] Thus, or so the dialectic goes, to force Dr, Knowles to participate in such practices would violate her integrity. Her integrity is something of great importance both to her and to others. It is important enough that in weighing the requests of the patient for MAID, an allowance might

---

[3] Stahl and Emanuel (2017).

[4] Consider the vast literature on referrals. See, for example, Aulisio et al. (2014), Greenblum, and Kasperbauer (2018) and Wicclair (2011).

[5] See Wicclair (2021).

[6] Sulmasy (2008).

[7] Sulmasy (2008).

be made to protect the conscience of Dr. Knowles and shift her care responsibilities in this case. To the contrary, CDs would have her leave the profession.

Cases like that of Dr. Knowles are common in discussions of conscience in medicine, but important features of these cases are not often emphasized. First, it may be that the deeply held moral beliefs of medical professionals ought to be respected, but if they are, it is because they are human beings for whom these things matter. Thus, there exists a tension between a CD's claims rooted in professional practices and norms and the integrity-based CP's focus on the humanity of the persons involved. This can lead to each talking past the other. In either case, to try to uncover and prove the existence of a professional's deeply held moral beliefs is fraught with challenges and practical attempts to do so are epistemically problematic. An impasse appears. Second, the role of religion is often cited in the standard set of cases, as the basis for the beliefs of those seeking protection. This attribution creates another, potentially deeper, impasse between those who think religious views are relevant to medicine (or, at least, should be respected) and those who do not. Third, the situation is often described as a conflict in which a practitioner's religious views are pitted against a patient's autonomous choice. Again, an impasse is found.

The approach followed in this text is nonstandard in part because it rejects many of these categorizations and assumptions. First, it diverges from accounts of conscience that takes seriously the need to identify and respect only deeply held beliefs. This is not because such respect is not important, it is; however, discerning in practice what it would be for a person to display the right behaviors and attitudes to indicate that a belief was sincerely held and deep is untenable. Thus, requests for employed health professionals to submit statements of belief for hospital ethicists to read and evaluate the veracity and depth of, so as to determine sincerity, are wrongheaded and epistemologically problematic. This practice is unethical for a number of reasons, foremost of which is that it is not a reliable method of ascertaining accurate information. The point here is similar to that made in research ethics contexts that research cannot be ethical if the scientific methodology being employed is not valid. There is no cost-benefit analysis that justifies burdens to participants when the science is faulty. The same goes for interrogation practices rooted in bad epistemology. Second, though religion is often cited as the flashpoint at the start of the fire that is the debate over CO in medicine; religion, in fact, has nothing to do with the debate over conscience. Even if religious views, beliefs, or practices are deeply meaningful to persons, it is not religion but conceptions of moral responsibility that matter in the case of conscience. Third, a number of the positions in this debate conceptualize persons as lists of values or as static sets of interests. This conceptualization might appear to aid in understanding the conflict, as it allows for more easily weighable claims, making commensurable what otherwise would not appear to be so; however, it is inaccurate. Persons are complex and they live rich moral lives, which are bound up in interesting but often challenging ways with other persons who are also trying to live lives similarly. This leads to a critique of the more standard positions in this debate: they fail to account for the rich moral lives of persons. Before returning to this analysis, a more detailed description of CPs and CDs is needed.

The imagery of poles has been useful[8] in describing the positions that one might adopt about the role of conscience in medicine. This is for a variety of reasons, but foremost among them are the complexity of human self-conceptions and the role that conscience might play in those self-conceptions (as well as the various disciplines, authorities or authoritative sources, and kinds of claims that are connected to considerations of conscience). One pole, toward which accounts of conscience in medicine gravitate, attracts *conscience dissenters*. CDs are proponents of positions that favor or argue for the primacy of professional standards, civil laws, or patient autonomy over the conscientious practice of individual professionals; that is, they argue for an external norm or, at least, a standard outside of the reflection of the individual practitioner. CDs might be contrasted with proponents of positions that give priority to the conscientious practice of individual practitioners, the aforementioned CPs. CPs argue for an internal norm, relying on and supporting the practical reasoning and judgements of individual practitioners. If it is right to describe positions in this debate as gravitating toward the poles of conscience protection and dissention, then it should be noted that there exist a variety of "middle of the road" and compromise positions that hover in the space between. Positions might reside in this middle area for any number of reasons. Some positions take care to distinguish those procedures or practices practitioners might define as outside of their conscientious practice and those that they may not. Depending on the arguments for the distinctions and which medical acts might be covered by exemptions for conscience (for example, procedures, specific referrals, general referrals, information-giving, et cetera), such a position might feel the pull toward one pole or the other. One might also offer a position rooted in the nuances of professional medical practice and the practicalities informed by on-the-ground training and care of patients. Such positions might allow conscience-related reasons as justifications under certain conditions; for example, one might allow professionals to place outside of their sphere of professional practice, certain procedures or practices as long as the relevant claims of conscience are shared in advance, communicated in certain ways, or do not cause undue burdens on others (where others could include both patients and fellow clinicians). Another set of positions which occupies a middle space is that which reject descriptions of appeals to conscience in healthcare as CO. Proponents of such views might do so for historical reasons, such as the idea that CO has a particular history bound up with wartime and pacifism and because the features associated with claims of conscience in medicine are distinct, a distinct term is needed. Proponents of such views might do so because what they see on the ground does not fit CO but is more akin to something like civil disobedience.

At this juncture, and picking up on this last point, let us distinguish conscientious practice from CO and civil disobedience.[9] For clarity, consider and compare the following cases: First consider a physician who objects to the practice of MAID. Suppose that they were trained in medicine, understanding it to be both art and science, and that they have read the Hippocratic corpus. They trace the history of

[8] Pilkington (2021).

[9] In her work on conscientious objection and civil disobedience, philosopher Howard (2021).

their profession to its Hippocratic roots and they understand it to, at its core, consist in a group of healers. The physician does not believe MAID is an act of healing, so much so that they balk at their colleagues' use of the term "medication" for the substances that are employed in MAID. The physician is an advocate for their profession and for the patients that they believe are owed the care of physicians, not the abandonment that they see in the prescribing MAID brings. They lobby their professional association, they pen op-eds in prominent newspapers about the subject matter, and they work to change minds about the acceptability of MAID. When MAID is legalized in the areas in which they practice, the physician redoubles their efforts to change minds and requests that patients seeking this service, at the hospital where they have privileges, not see them but be directed to another physician.

This case is an amalgam of real situations and captures a set of narratives of persons who find the practice of MAID to run against the ideals of Hippocratic medicine. However, this case is not intended to raise ethical reflection on MAID, but rather to highlight an instance of CO. A physician makes public claims, requests an exemption from their employer, actively seeks to reverse opinion on the subject—they are attempting to conscientiously object. Now whether they are successful or not is not the point, nor is whether they are granted an exemption and required to undertake certain penalties or sanctions, or to perform additional duties to "make up" for exempted work if granted.[10] The point is that they are publicly and actively disagreeing and requesting that space be afforded them to do so in both their practice and in their advocacy against MAID more generally.

We might distinguish this case from another wherein a different physician, who also finds MAID to violate the ethical requirements of medicine, actively blocks attempts to offer this service to patients. They stage a sit in of the hospital president's office. They lead peaceful protests in front of the hospital pharmacy where the MAID prescriptions are filled. They lead others to fill the hallway and obstruct persons attempting to fill the MAID orders. This physician appears to be civilly disobedient. They may conscientiously object, as well, were they pressed to provide this service, but that is not yet the case. They are obstructing others from providing and accessing MAID. They may be doing so in a polite and well promulgated manner, and in fitting with what they know about peaceful protests, and as such are being civilly disobedient.

Consider a third physician, who is trained in palliative care with a focus on end-of-life care. Supposed this third physician moves to a locale where MAID has recently been legalized and where, due to a feature of that location, it is in need of physicians who can offer this service to patients. The particular feature does not matter for the argument, it might be a rural area, for example, but the key is that it is under resourced. This physician, like the former two, finds MAID to be a violation of a physician's duties, and moved to a place where it had been legalized in order to reduce

[10] Thus, it may be the case that this example does not meet CO, as defined by Stahl and Emanuel. In fact, they might be correct that there is enough of a disanalogy between CO requests by pacifists in wartime and CO requests by physicians to think that they should not fall in the same category. However, this does not mean that physicians should be barred from conscientiously objecting, but rather that what is being requested is not CO. The difference is terminological, not normative.

the opportunities for MAID. This physician does not share their views in advance, nor do they (or will they) refer patients to other physicians for this service. They moved there to obstruct the expansion of MAID and to do so quietly, advocating change through making the service unavailable.

Consider a fourth and final case. Suppose a physician lives and practices in an area that recently underwent a legal change regarding the permissibility of MAID. They do not believe that physicians ought to engage in the practice but realize that others have different opinions about the matter, and they do not seek to change their views, to advocate or lobby for legal changes, nor to obstruct offices or pharmacies. Rather, they simply wish to continue practicing as they had, in line with their own understanding of what good medicine is, what health is, and how physicians might aid patients. This physician is not being civilly disobedient, like the second, and they are surely not being obstructionist like the third. Are they conscientiously objecting, like the first? The case could be described as such, noting that the hospital at which they practice might require them to testify to the possession of some core set of beliefs, and to holding those core beliefs deeply, in order that they not be forced to prescribe for MAID. However, this is not the best description of this situation. Examining the case carefully raises the question, "To what is the physician objecting?" The physician does not go in for MAID, but they are not protesting, seeking to make change, or even offering a general view on the subject. They would like to practice according to their own understanding of what good medicine is, that is, they would like to practice conscientiously.

These cases help to illustrate some of the terminological complexity in the debates over conscience and to support the suggestion that conscientious practice, not CO or civil disobedience, is central to this debate and is what relevant examinations of the subject matter should focus on. These cases also help to show how quiet (non-public) refusals to participate in a practice that a professional finds objectionable could be instances of conscientious practice or not. The professional who practices conscientiously and does not engage in CO or civil disobedience or obstructionist tactics is simply looking for space to practice without incursion and this is something different from, even if related to, those other concepts.

Physicians in each of the four cases do not support MAID, and the cases were written in that way with the purpose of honing in on features of those cases that might distinguish them from each other. However, it should also be noted that the content of the claim does not matter to the CP, as it may not matter for the CD (though that depends on the particular account). With respect to the CP, these four cases could have been written in such a way that the physicians were working toward a greater realization of MAID, and not the other way around. This is a crucial point to make for at least two reasons. First, it helps to clarify that advocates for conscience protection are advocates of individual persons, and they may have to bite some significant bullets about particular practices in order to support conscientious practice generally. Second, it prefigures a later discussion in this text about the role of religion in debates over conscience—a role that has been significantly overemphasized in the literature and in practice.

Thus, one could envision four structurally similar cases in a post-Dobbs era, wherein physicians engaged in all of the same actions (and omissions) but with respect to providing or advocating for greater access to reproductive services. The conceptual map would not change. Before moving to describe the disagreement between CPs and CDs, a discussion of the distinction between internal and external norms is merited, if for no other reason than to shed more light on what is meant by conscience and conscientious practice.

Recall that one important difference between CPs and CDs is their understanding of the place from which the relevant norms bear on the individual professional in question. CPs, focusing on the conscientious practice of individuals, allow for an internal norm—the conscience of the professional—to be operative. CDs, on the other hand, appeal to a norm that is external to the professional in question. This is the case even though proponents of different conscience dissenter positions appeal to different norms, e.g., professional standards, laws, patient autonomy or freedom, et cetera. It is important to note that highlighting this difference does not require advocating for an intuitionist account of conscience. It also does not suggest the view that a person's, or in this case a medical professional's, conscience is always right. Sulmasy's clarity on these points is instructive. He writes, in reference to intuitionistic accounts of conscience, in general, and in reference to Bishop Butler's claims, in particular:

> this is not an adequate theory of conscience. Conscience is not a little voice whispering to each of us infallibly about what we should do. Would that it were so simple. If conscience were nothing other than a sort of perfect perceptual intuitionism, all moral conversation would grind to a halt. The only thing moral philosophy could possibly say about morality would be "follow your conscience.[11]

With respect to the second point, about the rightness of the outputs of one's conscience, Sulmasy continues, writing:

> I am deeply skeptical about any form of act intuitionism as a theory of ethics. Our intuitions about particular cases will almost certainly differ. If they do, as they seem to in the troubling cases that confront us, such as abortion and physician-assisted suicide, then all we would be able to do would be to recognize that our intuitions differ. According to a theory of moral intuitionism, these differences could neither be explained nor challenged. This leaves open too many possibilities. My intuitions about what is right and what is wrong differ from those of the Janjaweed militia in Darfur. I want to reserve the right to challenge their intuitions.[12]

This is an equally critical point. Put to the side metaethical debates about moral intuitionism, as those are beyond the scope of the argument of this text. What matters here is two-fold. First, the status of the claims of conscience of one professional or another matters. Second, if disagreements occur within professions, outside of professions, or between professionals and non-professionals, how those disagreements might be resolved matters. Merely uttering "conscience" is insufficient. Such utterances ought not carry with them rhetorical power or stopping force without

[11] Sulmasy (2008).

[12] Sulmasy (2008).

clear argumentation. This is for any number of reasons, one of which is that claims of conscience can be wrong. Thus, that the norm is internal does not mean that one's conscience is "a distinct mental faculty, an intuitive moral sense that determines the rightness and wrongness of actions."[13] Rather, grounding a claim in one's conscience or citing one's conscience as a reason for particular actions comes from the individual, from something internal, as opposed to being placed upon one from the outside. This is relevant (especially in later chapters, where professions and disagreement are addressed) because in thinking about the conscientious practice of medical professionals and, more basically, what it means to be a professional and what a profession is, the latter is partially constituted by an ability of members of a group to hold off certain external influences.[14] When external regulation and control dominate a profession, it raises the question of whether the coordinated set of activities by its members still count as a profession. Thus, whether a norm that guides practice comes from an individual professional or comes, externally, from professional standards, legal codes, or the influence of others (e.g., patient autonomy movements) is an important distinguishing characteristic of CPs and CDs.

In highlighting this difference in norms, it is important to note that the potentially corrective nature of ethics (very broadly construed) need not be the sole claim of one of the parties in this debate. It may be tempting to think that the CDs have the upper moral hand because in appealing to something external, they capture morality or at least the potentially corrective nature of it. However, just as with consciences, laws and movements (even if both aimed at the good) can be mistaken. It is also not the case that consciences cannot incorporate the corrective feature of morality that appears so useful in these spaces. Consider the aforementioned definition of conscience "as having two interrelated parts: (1) a commitment to morality itself; to acting and choosing morally according to the best of one's ability, and (2) the activity of judging that an act one has done or about which one is deliberating would violate that commitment."[15] Thus, on some accounts of conscience, the corrective nature of morality is part of conscience.

In the end, whether CDs or CPs win the day will not be decided easily by an appeal to the corrective nature of morality. Rather, a deeper dive into the disagreements themselves is needed, which is the subject of the next section.

[13] Sulmasy (2008).

[14] Buchanan (2009).

[15] Sulmasy (2008).

1.1 The Debate Between Conscience Dissenters and Conscience Protectors

To take stock of the arguments of this chapter thus far, clusters of positions have been categorized primarily under two headings: conscience protectors and conscience dissenters. The former category contains a variety of notable proponents. Whether they would all count themselves as properly within this category is an interesting question, but the category includes physicians like Lauris Kaldjian,[16] philosophers like Mark Cherry,[17] and (arguably) bioethicists like Mark Wicclair.[18] In the conscience dissenter category are the likes of Stahl and Emanuel,[19] Julian Savulescu,[20] Jake Greenblum and T. J. Kasperbaur,[21] Udo Schuklenk,[22] and Ricardo Smalling.[23] A third cluster of positions, those middle of the road views, such as Wicclair (depending on one's interpretation)[24] or someone like Robert Card,[25] Paul Cummins,[26] or Thomas Harter[27] also exist.

Categories are well and good, but they need to do some work in order to justify the distinctions that demarcate them. Why does there exist deep disagreement about claims of conscience? That is, why do people care about it and what hangs on an appeal to internal norms as opposed to external norms? One answer to the first question[28] is that conscience, however it is conceptualized (CO, civil disobedience or conscientious practice, or simply considerations of conscience), is closely related to foundational issues in health, healthcare delivery, and institutions of healthcare. As bioethicist Nancy Berlinger writes, "Conscientious objection in health care always affects someone else's health or access to care because the refusal interrupts the delivery of health services. Therefore, conscientious objection in health care always has a social dimension."[29] In other words, considering the proper place of appeals to conscience in healthcare matters because it affects others and is bound up with something that many, many people value greatly: their health. In fact, this is one of the motivating ideas for the account offered later in this text, as well as its key insight: that conceptions of the moral responsibility are the key to unpacking disagreements about

---

[16] For his discussion of conscience, see Kaldjian (2014).

[17] For an example of why Cherry fits in this category, see: Cherry (2012).

[18] Though Wicclair's (2011).

[19] Stahl and Emanuel (2017).

[20] Savulescu (2006).

[21] Greenblum and Kasperbauer (2018).

[22] Schuklenk (2015).

[23] Schuklenk (2017).

[24] Wicclair (2011).

[25] Card (2020).

[26] Cummins (2021).

[27] Harter (2021).

[28] For a lengthy discussion of this topic see: Pilkington (2021).

[29] Berlinger (2023).

conscience. Thus, due to the interconnected nature of persons and the seriousness and intensity of their caring about their own and others' health, it should not be surprising that disagreements, and strong ones, arise. To the second question, it is worth noting that often (though not always) when norms are understood to be internal, individuals are shown deference with respect to their acceptance of and engagement with those norms; however, when norms are understood to be external, that is not the case. This may be why, in a piece intended to brief the public about conscience in medicine, Berlinger writes, "Conscientious objection in health care cannot be framed solely as an issue of individual rights or beliefs because it always affects someone else's health or access to care."[30]

Due to the fact that the focus of conscience, and the debates about it, are often directed at or connected to human features and not specifically medical in nature, the breadth of engagement is greater and the relevant questions are not merely reserved for those with particular medical training to answer. That is to say that views about the conscience in medicine may very well be connected to the place medical professionals hold or ought to hold in the delivery of care. Answers to questions like, "Who is a professional? What is a profession? How should professionals of different professions interact? How should professionals interact with institutions, those they serve, and members of society?" are needed in these discussions. Answers to these questions are connected to matters outside of medicine; that is to say that discussions of conscience in medicine are in some ways proxies for conversations about other matters, matters about which many people feel well placed to offer an opinion. This fact helps to explain why debates over conscience are fraught (at least as they are often engaged currently). It is because there is practical, conceptual, personal, and socio-cultural baggage that interlocutors bring to these debates and, as can be the case with baggage, it gets heavier as time goes on and it is not always noticed by others when one places it on the ground... though it is still there. This leads to the appearance that debates about conscience in medicine are intractable: commonly held packages of responses to innovative ideas, confusions over differing terminology, and proxy fights over what are really debates about foundational ethical matters have found their way into debates over conscience.

One reason that the debate can be so perplexing is that though it might be recognized as intractable, because it is connected to something so prized, health, it is not a debate that can be ignored. This is, in part, because claims of conscience can affect the daily tasks of healthcare, thus impacting professionals and the patients (and families) that they serve. These claims also impact institutions and, as a result of both, they influence the communities of which professionals and patients are a part and where institutions exist. Though many of the debates involve philosophical questions and arguments, philosophy in this space very quickly leads to practical questions and issues. Take a seemingly straightforward philosophical question about conscience: Should medical professionals be allowed to practice conscientiously, that is, should they be afforded the personal or moral space to act in line with the dictates of their conscience? A question like this admits of a variety of philosophical answers rooted

---

[30] Berlinger (2023).

in theoretical commitments to conceptions of conscience, ethical theories describing what relationships between persons ought to be like, et cetera; and, yet, these theoretical answers impact very practical concerns. It is a short step from these philosophical considerations to questions, such as "What are the logistical steps needed to address a vacancy (or an addition) in care in an instance where a professional's conscientious practice does not include (or does include) kinds of care that the institution, of which the professional is a part, offers (or does not offer)? Consider also theoretical commitments tied to accounts of interpersonal connectedness in action theory. Reflection on this set of philosophical considerations which quickly leads to practical questions, such as "What are the consequences of referral policies at a particular institution for members of particular communities? Such practical questions call out for responses that are both consistent with theoretical commitments and, in this case, the relevant ethical theory. Further, they must also take into account the kinds of logistical issues mentioned above, e.g.: What are the potential burdens on professionals on the same shift as the one who is conscientiously practicing? Are human resource policies in line with or opposed to the existence of conscientious practice? Does the institutional mission rule in or rule out particular forms or features of conscientious practice in a way that such practice would lead to deep confusion regarding the expectations of reasonable members of the community that the institution serves?

The preceding framing sets the stage for the debate over conscience. As the reader works through the arguments in later chapters, deciding for themselves which position or cluster of positions is most defensible, four general questions arise.[31] These questions should be considered with two caveats in mind. First these questions are asked in the service of the more general approach in this text, which takes a step back from the current debate and asks whether conscience claims (and, in turn, conscientious practice) can be defended? Second, they are asked with the presumption that generally agreed upon resolutions to a number of the issues connected to conscience do not yet exist (which is a contentious claim). This is, in part, because of the manner in which debates over conscience have been framed. In such debates, descriptions of appeals to conscience, claims of conscience, invocations of conscience clauses, conscientious objections, being civilly disobedient for conscience-based reasons, and the like are all offered under a narrative in which an individual professional (or set of professionals) exists in opposition to a larger group or an institution. This description of the one (or the few) against the many takes a variety of forms. Sometimes cases involving appeals to conscience include a subset of members of a profession against the agreed upon norms of a larger professional association (which usually means a greater number of members or of other professionals). Other times, cases involving appeals to conscience involve individuals within a wider cluster of professionals, say, within a community of practice,[32] against other professionals within a distinct community of practice (e.g., consider a debate over covid era hospital visitation policies that involved physicians and nurses, on the one hand, and social workers and patient advocates on the other). For a third example, consider persons seeking

---

[31] For an earlier discussion of these questions, see: Pilkington, Bryan (2021).

[32] For more on communities of practice and their benefits, see: Delgado et al. (2021).

specific healthcare services against the backdrop of a health system that was reticent to offer such services. What these narratives, which are likely familiar to the reader, illustrate is an oppositional framing. It should be noted that the argument is not for the conclusion that these cases do not exist nor that they should all be reframed in a more congenial narrative. Rather, it is that the narratives cohere around an important truth about appeals to conscience in medicine and debates over those appeals, yet they do not tell the full story. Instead, these narratives offer a superficial gloss, obscuring similarities between the parties to the disagreement. Part of what is obscured is the rich moral lives of persons in all places within these narratives. What is also obscured is the role that conceptions of moral responsibility have in all perspectives within these debates. This means that different accounts of conscience and its place in medicine,[33] which will be surveyed later in this text, will have to address philosophical (not only practical) questions. These questions include: How does a particular account of conscience adjudicate between competing claims? How does a particular conceptual framework involving a place for conscience place interlocutors in the best situation to resolve disagreement? Where and how is this done? With this preface and these qualifications in mind, consider the four aforementioned questions.

First, it is important in surveying the relevant positions to ask what is the proper role of conscience (and appeals to conscience) in medical practice (and healthcare broadly)? Relatedly, it should be asked what terms are (and ought to be) employed in discussing, reflecting on, and evaluating that role. It is not uncommon for accounts of conscience in healthcare to allow for some space for professionals to make judgements about what is appropriate practice relative to a health-related matter for which they are being consulted. What varies drastically is the amount of space that is afforded and the kinds of reasons or the types of judgment that are permitted to occupy that space. For example, the arguments of this text support a fairly wide space for professionals to exercise their best judgment relative to their own conceptions of moral responsibility. This is not a wide space, in general, but rather it is focused on that set of conceptions. Other accounts, even some offered by CDs, allow

[33] This is no small claim. Various of accounts of conscience with different emphases and steeped in philosophical analysis to different degrees can be illustrated by considering any number of them. One influential argument in favor of conceptualizing appeals to conscience in medicine as conscientious objection and then allowing for them comes from the clinical ethicist Harter (2021). Harter argues that critiques of conscientious objection are often arrived at too hastily, fail to appreciate the practical side of conscientious objection and possibility of practitioners' perspectives changing. He claims that many of the standard arguments against conscientious objection do not apply in all cases, leading to the conclusion that an opportunity for conscientious objection should be tolerated but in a limited set of cases. Other accounts highlight the arrays of influences on a professional's conscience, such as the work of Fleming and Ramsayer (2021) which makes clear the impact of professional codes, state legislation and, in an interesting case they take up from Scotland about midwifery, the dictates of church teaching. Other work on conscience in medicine highlights the role and employers play in considering claims of conscience. As Paul Cummins has argued (Paul 2021) in describing his relational account involving employed physicians. Even though this short list of accounts, intended to illustrate the variety of positions in the conscience debate, has focused on appeals of conscience in terms of conscientious objection, not all positions do. See Howard (2021). For an argument that civil disobedience is a more fitting description for many instances of appeals to conscience.

space for the judgement of professionals. For example, Stahl and Emanuel allow for medical professionals to make determinations about issues that have not been "settled" yet within professions.[34] Thus, they allow for professionals to make their own determinations about, for instance, MAID. However, any "settled" matter is not something about which professionals may exercise judgment.[35] Other CDs, consider Greenblum and Kasperbauer,[36] appear to offer no place for physician judgement, as they understand medical professionals to be semi-public employees. This is the case unless such judgment is requested by patients, thus, yielding a theoretically possible but practically unlikely space for said judgment. There are also positions, such as those offered by Savulescu,[37] which simply rule out space for conscience.

It should also be noted, partly in answer to the question about terminology, that this use of "space" may be overly vague, and this use of judgement may be contentious. I ask the reader's indulgence on the first; a good imagination should do sufficient work to resolve any concerns there. With respect to the second, there is much to be said, and the idea will be given more attention later on in the text. This use of judgement is contentious because there are all kinds of judgments in medicine and CDs distinguish (though often implicitly) between moral or ethical judgments and clinical ones. This is a fair and important distinction, but it is misused. There are cases in which the lines blur and what a professional is doing when they make judgments may be a bit of both. Moral and clinical components of medical judgements are more intertwined than is sometimes admitted. CDs put a great deal of stock in the distinction because that is what allows for arguments in favor of professionals as semi-public employees or the role of professional associations as dominant or claims like "one ought to leave one's religion at the door." To be clear, I am not arguing that clinical judgements and moral judgments are the one and the same; I am pointing out that this area is murkier than some suggest and to think that one does not influence the other is misguided and naïve.[38] It is, however, part of the conscience dissenter narrative: if there is a moral dimension to clinical judgements, if moral reasoning and clinical reasoning are intertwined or mutually-influencing, then it becomes much more challenging to object to conscientious practice. There are, to be sure, benefits to such a position, not only costs (or perceived costs). The logistics of patient care and the administration of hospital and related matters do become more complicated. The value-laden nature of clinical reasoning is not always admitted to or fully understood, but it is at the heart of exactly these kinds of disagreements. Discussions of conscience often involve

34 Stahl (2017).

35 This view is problematic because of its inability to make sense of progress and change with disciplines. Though critiques of this position are taken up later in the text, it is referenced here to highlight a similarity across positions in the debate: many allow for some space of the aforementioned kind; the question is how much and of what type.

36 Greenblum and Kasperbauer (2018).

37 Savulescu (2006).

38 This is not a new critique. Similar lines of argument exist in discussions of liberalism in political theory, in particular in work on public reason. It is thus no surprise that the work of Rawls (1999, 2005) is often referenced in these discussions and informs the kind of RE invoked by Stahl and Emanuel.

value or values-laden language, as well, and—even if misguided and naïve—are seen by some as involving personal rather than professional interests. Thus, again, it is important to take care to notice terminology that is employed in arguments offered by proponents of conflicting positions in these debates. Consider the commonly used term "conscientious objection." Though it does not capture what is often going on in instances under scrutiny, as argued above, it appears to be the preferred term for many bioethicists working on the subject matter.[39] Yet, this should not be surprising given the aforementioned discussion of how the issue of conscience in medicine is often framed as an adversarial one. To the point, one set of CDs analogizes disagreements about conscience in medicine with political disagreements over contributions in war efforts, as noted above[40] to argue against the permissibility of appealing to one's conscience in medicine.[41] Still the term "conscientious objection" brings with it its own conceptual and rhetorical baggage, just as any term does. Whether a term like "civil disobedience" or "conscientious practice" better captures what is going on in the cases under consideration will be borne out in the arguments, which is not to say that a clear preference is not expressed in this text.

The preceding and earlier discussions of terminology aimed to illustrate that instances referred to as CO occur within specific contexts, against a backdrop of clearly delineated scopes of practices, and as defined by professions (and licensed by authorities). It is in this kind of environment that those who have (what are viewed to be) different conceptions of their own practice are labeled "conscientious objectors." One line of reasoning which prefers the term conscientious practice over CO is rooted in applicability. The term "conscientious objection" does not apply as widely to practices in medicine as CDs assume. Further, within a discussion of conscientious practice, the terminology of "conscience protectors" and "conscience dissenters" is preferable because of the failure of CDs to offer a clear and accurate picture of the essential contextual features relied on in discussing conscience. A particularly challenging misunderstanding for CDs to overcome is the failure to correctly account for what a profession is and explain how disagreement is to be resolved within such an organizational unit. Descriptions make a difference, and no terminology is free from connotation: describing one position as a protector and the other as the dissenter is not by accident nor are these rhetorically neutral terms. The reader ought to attend to terminology, including its connotations, and the rhetorical power it might carry. I now turn to a second question that should be kept in mind when evaluating these arguments: Who are the parties involved, that is, who are the stakeholders?

This second question is useful because endeavoring to answer it may assist in uncovering those who might benefit from or bear the costs of the allowance of conscientious practice. Trying to understand who the relevant stakeholders are might aid in highlighting where power exists and what interests are invested in particular answers to questions over conscience. Many times, discussions of conscience (regardless of

---

[39] See, for example: Savulescu (2006), Schuklenk (2015), Schuklenk et al. (2017), and Stahl (2017).

[40] Stahl et al. (2017).

[41] See also Genuis (2021), Kane (2021), McConnell (2021).

whether one is a CP, a CD, or holding a middle position of some stripe) focus on the relationship between the patient and the physician. This is a sensible focus given the history of that relationship in medicine and the significant amounts of ethical reflection on that relationship. However, to do so exclusively is to miss important features of the current contexts in which conscience is appealed to. To focus on the physician-patient relationship in cases of conscience might lead some, as it led Greenblum & Kasperbaur,[42] to attend to power dynamics within that relationship and potential vulnerabilities that might be exploited or not responded to in a fitting manner. These are important facets to attend to, but they are not the only ones. In fact, this dyad does not capture the complex and multifaceted arena of healthcare delivery currently. As Paul Cummins argues,[43] the relationship is three-fold and must also include the physician's employer. If this is right, and Cummins arguments are persuasive, then the employed status of a physician should make a difference in our theorizing about conscience in medicine. Cummins highlights that respecting a physician's conscientious practice or accepting a CO request is, then, highly context dependent for what are, in fact, physician-employees. Cummins' argument also makes clear that the power dynamics are more complicated and medical professionals, as well as patients, are vulnerable, though in distinct ways. Cummins' analysis of physician-employees is more nuanced and more robustly argued for than the claims of Greenblum & Kasperbaur about physicians as semi-public officials; however, reflection on both analyses illustrates that part of the outcome of the arguments over conscience in medicine hang on what constitutes a profession. That notion, and the third question, are considered next.

What is the role of professions within determinations of the acceptability of appeals to conscience in medicine and how ought that role be assessed? Should members of a sovereign profession be afforded, among their special privileges, opportunities to practice conscientiously?[44] That this debate involves professions (and professionals, in a strict sense) matters. If medical professionals are understood to simply be possessors of an occupation, not members of a profession, the kinds of professional autonomy, privilege, and deference that are granted to such members would not be in place to motivate the positions in the debate or to generate some of the tensions that give rise to the disagreements around conscience in medicine. Some argue that it is not professions or the concept of professionals that should do the normative work in this context, but rather social[45] or legal norms.[46] Others root

---

[42] Greenblum, and Kasperbauer (2018).

[43] Cummins (2021).

[44] Referring to physicians as members of a "sovereign profession" is a reference to the subtitle of Paul Starr's Pulitzer-prizing winning work in medical sociology (Starr 2017). Some disagree with the relevance of Starr's work on the origins of the medical profession in the United States. This is because the special status that physicians carved out for themselves, including the special privileges and power that come with such standing is not be as strongly felt as it once was. However, as the ethical arguments about the role of conscience in medicine must engage conceptions of professions, the early history of the medical profession is both useful and relevant.

[45] See, for example: Schuklenk (2015).

[46] See, for example: Schuklenk (2017).

their analysis of the role of conscience in medicine from within professions[47] but arrive at a very different analysis than that offered in this text. In contrast, bricklayers and window-hangers do not usually try to avail themselves of the protections of conscience (though, interestingly enough, they usually can refuse to take a job which is one of the very items at issue in the debate over conscience in medicine). There is much to say about professions, in general, and the medical profession, in particular, which is addressed at length later on in the text. Before concluding this chapter, a final question about the positions in this debate, one which follows from consideration of the various kinds of norms that might be relied upon in assessing the role of conscience in medicine: disagreement, must be taken up.

A fourth question, which is really a cluster of questions, centers on the notion of disagreement: Where does it exist, and between whom? How might it be resolved? What resources do different accounts of conscience in medicine bring to bear on potential resolutions? Disagreement exists, as noted above, in any space where there are conflicting views, and it can be especially challenging to resolve when the relevant stakeholders and parties to the disagreement care so deeply about what is at stake. These two features are strongly represented in the debate over conscience in medicine as medical professionals care deeply about their own health, the health of persons under their care, and their place in that care. Some theorists attempt to resolve disagreement through democratic or democratic adjacent processes. For example, Stahl and Emanuel rely on Rawlsian reflective equilibrium (RE) to adjudicate between intraprofessional disagreement, though it is argued later that their attempt fails. Others rely on philosophical distinctions in an attempt to do justice to what are seen as both sides of an argument, both with merit. For example, the distinction between "traditional claims of conscience" (the opportunity for professionals to not be involved in practices or procedures they find objectionable) and "contemporary claims of conscience" (the opportunity for professionals not to be connected in certain ways to practices or procedures they find objectionable) has been offered as a solution to this disagreement.[48] This proposal, which has been best articulated by Aulisio and Arora, suggests that practitioners be allowed space to not perform certain practices, but that patients could request direct referrals[49] to practitioners who do engage in the practices in question. This work aims to respect the integrity[50] of the medical professional and also satisfy the care needs of patients.

With these questions in mind, let us turn to a deeper discussion of the issues of disagreement that plague accounts of conscience in medicine.

---

[47] See, for example: Stahl (2017). However, as will be argued later, these authors really have in mind professional associations.

[48] Aulisio et al. (2014).

[49] The literature on referrals is extensive; for two influential arguments, see Wicclair (2011) and also Wildes (1993).

[50] Respect for the integrity of professionals carries weight with many CPs and middle view advocates, including, Aulisio and Arora, Mark Wicclair (depending on one's interpretation), and Jorge Garcia (see Aulisio et al. 2020); for a book length treatment, see: Kaldjian (2014).

1.2 Conclusion

In this first chapter, portions of the groundwork for discussions of key issues related to conscience that are more fully addressed in the subsequent chapters have been laid. Though a full accounting of all potential positions on the subject matter is not contained in this chapter, key positions have been mentioned and, more importantly, the most significant clusters of positions in the debates– those of CPs and of CDs—have been described. In this chapter, it was argued that positions gravitate toward either of those two poles with a variety of compromise positions floating in between. At their essence, CPs give primacy of place to the consciences of individual practitioners. At their essence, CDs give primacy of place to some other normative standard, be it legal norms, social norms, professional norms, or the autonomy of patients. In this chapter, pains have been taken to make the point that language matters and that appeals to conscience or conscientious practices are sometimes referred to as "conscientious objection," sometimes thought of as existing under the heading of civil disobedience, sometimes simply known as conscience (in medicine). The preferred term should be conscientious practice because it casts a wide net and captures the relevant issues and positions in the debate. Finally, this chapter—if its analyses are sensible, its readings of relevant claims reasonable, and its arguments sound—has built the foundation for serious discussions of disagreement that exists around conscience (Chap. 2), in particular of disagreement around the role of religion in appeals to conscience in medicine (Chap. 3), of moral responsibility (in Chap. 4), and, finally, of a medical act (in Chap. 5). It is to disagreement that we now turn.

References

Aulisio, Mark P., and Kavita Shah Arora. 2014. "Speak No Evil? Conscience and the Duty to Inform, Refer or Transfer Care." *HEC Forum: An Interdisciplinary Journal on Hospitals' Ethical and Legal Issues* 26 (3): 257–266. https://doi.org/10.1007/s10730-014-9242-8.

Aulisio, Arora, Mark Wicclair, and Garcia Jorge. 2020. "Virtues and principles in biomedical ethics." *The Journal of Medicine and Philosophy* 45 (4–5): 471 503. https://doi.org/10.1093/jmp/jhaa013

Berlinger, Nancy. 2023. "Conscience Clauses, Health Care Providers, and Parents." The Hastings Center. https://www.thehastingscenter.org/briefingbook/conscience-clauses-health-care-providers-and-parents/.

Buchanan, Allen E. 2009. *Justice and Health Care: Selected Essays*. New York: Oxford University Press.

Card, Robert F. 2020. *A New Theory of Conscientious Objection in Medicine: Justification and Reasonability*. Routledge Annals of Bioethics 21. New York: Routledg.

Cherry, M. J. 2012. "Conscience Clauses, the Refusal to Treat, and Civil Disobedience—Practicing Medicine as a Christian in a Hostile Secular Moral Space." *Christian Bioethics* 18 (1): 1–14. https://doi.org/10.1093/cb/cbs009.

Cummins, Paul J. 2021. "Conscientious Objection and Physician—Employees." *HEC Forum* 33 (3): 247–268. https://doi.org/10.1007/s10730-019-09390-8.

Delgado, Janet, Serena Siow, Janet de Groot, Brienne McLane, and Margot Hedlin. 2021. "Towards Collective Moral Resilience: The Potential of Communities of Practice During the COVID-19

Pandemic and Beyond." *Journal of Medical Ethics*. https://doi.org/10.1136/medethics-2020-106764.

Fleming, Frith, and Ramsayer. 2021. "Tensions Between Ethics and the Law: Examination of a Legal Case by Two Midwives Invoking a Conscientious Objection to Abortion in Scotland." *HEC Forum* 33 (3): 189–213. https://doi.org/10.1007/s10730-019-09378-4.

Genuis, Quentin I T. 2021. "A Genealogy of Autonomy: Freedom, Paternalism, and the Future of the Doctor—Patient Relationship." *The Journal of Medicine and Philosophy: A Forum for Bioethics and Philosophy of Medicine* 46 (3): 330–349. https://doi.org/10.1093/jmp/jhab004.

Greenblum, Jake, and T. J. Kasperbauer. 2018. "Forget Evil: Autonomy, the Physician–Patient Relationship, and the Duty to Refer." *Journal of Bioethical Inquiry* 15 (3): 313–317. https://doi.org/10.1007/s11673-018-9854-9.

Thomas Harter. 2021. "Why Tolerate Conscientious Objections in Medicine." *HEC Forum* 33 (3): 175–188. https://doi.org/10.1007/s10730-019-09381-9.

Howard, Dana. 2021. "Civil Disobedience, Not Merely Conscientious Objection, In Medicine." *HEC Forum* 33 (3): 215–232. https://doi.org/10.1007/s10730-020-09417-5.

Kaldjian, L. C. (2014). *Practicing Medicine and Ethics: Integrating Wisdom, Conscience, and Goals of Care*. Cambridge, U.K: Cambridge University Press.

Kane, Tiernan B. 2021. "Reevaluating Conscience Clauses." *The Journal of Medicine and Philosophy: A Forum for Bioethics and Philosophy of Medicine* 46 (3): 297–312. https://doi.org/10.1093/jmp/jhab005.

McConnell, Doug. 2021. "Conscientious Objection in Health Care: Pinning down the Reasonability View." *The Journal of Medicine and Philosophy: A Forum for Bioethics and Philosophy of Medicine* 46 (1): 37–57. https://doi.org/10.1093/jmp/jhaa029.

Pilkington, Bryan. 2021. "Considerations of Conscience." *HEC Forum* 33 (3): 165–74. https://doi.org/10.1007/s10730-021-09457-5.

Rawls, John. 1999. *A Theory of Justice*. Rev. ed. Cambridge, Mass: Belknap Press of Harvard University Press.

Rawls, John. 2005. *Political Liberalism*. Expanded ed. Columbia Classics in Philosophy. New York: Columbia University Press.

Savulescu, Julian. 2006. "Conscientious Objection in Medicine." *BMJ* 332 (7536): 294–297. https://doi.org/10.1136/bmj.332.7536.294.

Schuklenk, Udo. 2015. "Conscientious Objection in Medicine: Private Ideological Convictions Must Not Supercede Public Service Obligations." *Bioethics* 29 (5): ii–iii. https://doi.org/10.1111/bioe.12167.

Schuklenk, Udo, and Ricardo Smalling. 2017. "Why Medical Professionals Have No Moral Claim to Conscientious Objection Accommodation in Liberal Democracies." *Journal of Medical Ethics* 43 (4): 234–240. https://doi.org/10.1136/medethics-2016-103560.

Stahl, Ronit Y., and Ezekiel J. Emanuel. 1979. "Physicians, Not Conscripts—Conscientious Objection in Health Care." *New England Journal of Medicine* 376 (14): 1380–1385. https://doi.org/10.1056/NEJMsb1612472.

Starr Paul. 2017. *The Social Transformation of American Medicine: The Rise of a Sovereign Profession & the Making of a Vast Industry: With a New Epilogue by the Author*. Updated edition. New York: Basic Books.

Sulmasy, Daniel P. 2008. "What Is Conscience and Why Is Respect for It so Important?" *Theoretical Medicine and Bioethics* 29 (3): 135–149. https://doi.org/10.1007/s11017-008-9072-2.

Wicclair, Mark R. 2011. *Conscientious Objection in Health Care: An Ethical Analysis*. Cambridge: Cambridge University Press.

Wicclair, Mark R. 2013. "Positive Claims of Conscience and Objections to Immigration Law." *AMA Journal of Ethics* 15 (3): 188–192. https://doi.org/10.1001/virtualmentor.2013.15.3.ecas1-1303.

Wicclair, Mark R. 2021. "Commentary: Special Issue on Conscientious Objection." *HEC Forum* 33 (3): 307–324. https://doi.org/10.1007/s10730-021-09458-4.

Wildes, K. (1993). "Conscience, Referral, and Physician Assisted Suicide." *The Journal of Medicine and Philosophy* 18 (3): 323–328. https://doi.org/10.1093/jmp/18.3.323.

Chapter 2
Professional Disagreement

This chapter addresses the topic of disagreement among healthcare professionals. In the preceding chapter, part of the landscape of the debate over conscience in medicine was laid out. Though every position relevant to the discussion was not captured, two camps (or poles toward which proponents of particular views gravitate) were described: conscience protectors and conscience dissenters. CPs, as noted above, give priority to the conscientious practice of individual practitioners and CDs give priority to something external to the practitioner, such as professional standards, civil laws, and patient autonomy. Part of the tension that arises between these two positions has to do with the role that professions play in this debate. Some argue that professional associations are the proper arena to address disagreements. Others root their critique of conscientious practice in the professional standards of healthcare practitioners. Still others critique professions as the entities or professional associations as the groups or places wherein the norms that ought to govern individual appeals to conscience. What is evident from Chap. 1 is that professions must be given a serious look in any treatment of conscience, and this is what is done in this chapter as a way of attending to these disagreements. To do this, the first section considers what a healthcare profession (and derivatively, what a healthcare professional) is, borrowing from influential work in sociology and in the history and current practice of medicine. After briefly suggesting what the implications of these (and related) accounts on professionalism are, the existential question is raised: do healthcare professions exist? Next, considerations of disagreement in practice, and then of the common good and moral diversity, are taken up. The chapter concludes with a discussion of where the burden of proof lies in the overall debate.

© The Author(s), under exclusive license to Springer Nature Switzerland AG 2025

B. Pilkington, *The Medical Act: Conscientious Practice in a World of Dissention and Disagreement*, The International Library of Bioethics 114,
https://doi.org/10.1007/978-3-032-00048-4_2

2.1 What Is a Healthcare Profession?

There are at least four reasons that a discussion of professions is necessary for a proper treatment of conscience in medicine. First, given the importance that identification as a member of a profession holds for medical professionals, professions ought to be addressed. Second, the prevalence of appeals to professional spaces as places to adjudicate disagreement connected with issues of conscience also recommends addressing professions. Third, the norms that arise for physicians qua medical professional and the role that those norms play in discussions of conscience suggest a discussion of professions. Fourth, the significant role that professionalism and professional identity formation has begun and continues to play in medical education adds to the need to address professions. Claims rooted in professionalism are or, at least, ought to be dependent upon one's account of professions. Though a conception of professionalism could float free from such moorings, this is not the argument offered in this text. This leads us to the first key question entertained in this chapter. What is a profession? Though there are a variety of articulations, professions are generally understood to be associated with a set of relatively complex tasks performed by persons for the good of others. The skills needed and the altruism required are what set professions apart from mere occupations or hobbies. They are set apart in at least two crucial ways. The work of Allen Buchanan is instructive here. Buchanan discusses two conceptions of a profession: the ideal conception and the sociological conception. The ideal conception is a list of five necessary and sufficient conditions for a thing to be a profession. The sociological conception appends a sixth feature, but really it is a point about the ontological status of professions. It is to these conceptions that we now turn.

For Buchanan, and for the arguments in this book, what makes something a profession is that: first, its practitioners exhibit a "special knowledge of a practical sort." For example, physicians know how human bodies work and they have the skill set to diagnose certain ailments and recommend certain treatments. Not all persons have the knowledge that physicians possess, and it is (at least often) actionable, so their knowledge is both special and practical. Considering a medical specialty will also illustrate this point. Surgeons, for example, have a practical skill set—they are able to cut and stitch with very fine precision, as well as being in possession of a very detailed knowledge of human anatomy—and with their practical reasoning, they put both together. Not all persons can cut and stitch well nor do they have a similarly deep knowledge of human anatomy and the innerworkings of the human body. This example illustrates a knowledge possessed by professionals (and a subset of those professionals) that is both special and practical.

Second, professions (or members of professions) must have a commitment to preserving and enhancing the special knowledge that they possess. Professions, or their members, are charged with sustaining the knowledge base of their field and improving it. It is important to emphasize here that the special knowledge is of a practical sort. Professions are not, like those medieval communities residing in monasteries, endeavoring to safeguard a body of knowledge; collecting the information is insufficient to satisfy this requirement. This qualification might be (in part) satisfied by professionals sharing the latest techniques and discussing issues in the field (and related subfields) with other professionals. In medicine, this kind of engagement with other members of the profession is evidenced by the existence of such things as peer-reviewed journals, conferences, and grand rounds. Add to this work of enhancing special knowledge the education of future members of the profession. Professional educational institutions, such as medical schools, aid in the passing on of relevant knowledge to those who will become members of the profession. With these features in place, the requirement is satisfied, as the body of knowledge with a practical aim is passed on, and in a way that it might be put into practice.

The third qualification that must be satisfied is that members of a profession (and, arguably, the profession itself) must be committed to achieving excellence.[1] This requirement is relatively straightforward and, for many persons engaging these topics, it is intuitive. For medical professionals, as well as other health professionals and even professionals generally, executing the tasks that one is set to perform and executing them well are a condition for one to be a practitioner of that craft. This often requires practice (repetition of particular tasks or sub-tasks), the humility to be trained in particular methods (though not always), the willingness to stay up to date on the latest knowledge enhancements in the field (see the second condition), and the discipline to act and not act in certain kinds of ways (which might put practitioners in a place where the excellence of their practice is threatened). Though these ideas may be intuitive, consider three examples to illustrate this condition more fully.

First, consider a tennis player who wakes up early and spends a large portion of the day on the tennis court in training sessions and practice matches. They accept revisions to their style of play by coaches, tweak the trajectory of their swing, and study alternative approaches and styles of play. They consider ways to combat particular moves on the court and responses to those responses. When finished, they spend time lifting weights, doing agility training, running, and eating meals which are tailored to produce optimal athletic results. They do this most days of the week, most weeks of the year, for the duration of their careers.

Second, consider a teacher who wakes up early to prepare lessons for the day and remind themself of the individual strengths and weaknesses (and even preferences!) of their young learners. Then, the teacher prepares the classroom, teaches all day, and sets things up for the next day. They attend to the latest advances in learning theory and revise their curricular endeavors each year. They do this every weekday, throughout most of the year, and for the duration of their time as a teacher.

Third, consider a physician who, like the other two practitioners, also wakes up early. They see patients, round with residents, and stay late to ensure that a few of

[1] The focus in most descriptions of this requirement, including Buchanan's work, is on the aims of members of the professions. This is correct. However, the profession is added here, as there may be broader policies, practices, and activities which aid (or force) members to achieve (or at least strive for) excellence. Consider profession-wide recommendations about time off or work hour limits as a means to positively impact the wellbeing of professionals and to contribute to the achievement of excellence in their work product.

the more challenging issues facing their patients are managed well. In the evening, they catch upon the latest advancements in their field by reviewing recent journal articles, before working on a draft of their next conference presentation which they aim to publish in one of those aforementioned journals. They do this daily, excepting their day off, for the duration of their time caring for patients.

To achieve excellence (or to be in the process of striving for excellence) in any of these three pursuits requires practice: one must swing a racquet, interact with learners, and take a history and physical many, many times in order to get it right and, eventually, to do it well. One must be sufficiently humble to work under the tutelage of another—coach, senior educator or teacher, attending physician or chair (depending on the stage one is at). One must know what is going on in one's field: what are the latest advances in knowledge, what other members are doing, how other professionals have solved shared problems, reduced costs, and increased benefits, et cetera. One must also avoid situations that threaten that excellent practice: a lack of sleep and poor food choices on the eve of a match might be harmful to playing well, not taking sufficient time for rest may lead to a reduction in the capacity for patience that is crucial for educating the young, excessive drinking or use of recreational drugs before seeing patients could lead to poor judgment and the failure to diagnose and offer treatment plans accurately.

The fourth condition, which is common to most (if not all) descriptions of professions, is altruism. In Buchanan's formulation, the requirement of altruism is articulated as an "intrinsic and dominant commitment to serving others." This is an important condition in that it rules out many occupations as professions or, at least, requires more significant argumentation and a broader allowance for what constitutes altruistic behavior were many occupations to count as professions. Consider the aforementioned examples. The medical profession houses, or is supposed to house, members who possess that commitment to serving others. Medical professionals' primary obligation is to care for others. They are healers and the ethic(s), at least in the rituals that mark the beginning and conclusion of formal medical education, find professionals reciting the Hippocratic Oath or an updated version of it, such as the Physician's Pledge. When physicians are understood to no longer be working from altruistic motives (putting to the side mistakes, failures, and the bad actors which exist in any group or endeavor) the deference that is offered to them is restricted. Another illustration of the necessity of the altruism built into the profession, at least as it is realized in certain locales and contexts, are the legal prohibitions on walking out, striking, or other labor negotiation strategies and practices. Patients and their care are primary; to satisfy this altruistic condition is to put others before oneself. This brings us to the fifth condition, which is about self-regulation, and the subsequent conception (an additional sixth condition) of what a profession is, but first contrasting medical professionals with professional athletes and teachers will prove instructive.

One might make the case that teachers are members of a profession, surely, they fit some of the criteria and the kind of education offered—especially in the early years of a student's life—may highlight a dominant and intrinsic commitment to serving others. It is not the aim of this section of this text to argue against other pursuits being added to the list of professions. Classically, medicine, law, and ministry were understood to be the three professions, but given the existence of new occupations and the more careful theoretical work on what constitutes a profession, surely the list is longer. What might keep teachers off the list, on one interpretation of Buchanan's criteria, has to do with the objects of the professional's attention. Teachers, even if altruistically motivated and obligated in ways that satisfy Buchanan's conditions, are oriented toward their students. They do not (or, at least, an argument would need to be offered as to why they do) owe things to other students qua teachers. The teachers at my daughter's preschool are excellent in all of the ways one might hope for and satisfy these conditions, but they do not owe a good lesson or their best educational plans to students at other schools. Physicians are different. The title of the Buchanan's work in this space asks whether there is a medical profession in the house? Alluding to the old adage, is there a doctor in the house? If physicians are on a plane headed for a vacation and another passenger's health is in peril, they are expected to assist. If a physician is shopping for groceries and the grocer falls over clutching their chest, the physician is expected to assist. Again, as the aim of this section is not to rule out potential professions but to offer an account of professions to better inform a discussion of disagreement within professions and of conscience, whether teachers count or not is outside the scope of this book. Also, outside the scope of this book is whether professional athletes, if they give great joy to those who watch their sporting events, ought to count as members of a profession, having satisfied some account of being altruistic. There are other arguments, related to compensation, practices, and the institution of professional tennis (or professional baseball, soccer, et cetera) which suggest against identification with an intrinsic and dominant commitment to serving others; but it is enough, for the argument of this section, that physicians count as professionals. The other examples are offered to highlight potential similarities and instances of contrast.

The final condition of Buchanan's account of professions is the effective collective self-regulation and sanctions of professionals. If something is to be a profession, in light of the expertise of its members, their shared aims, and similar practices, it ought to be able to regulate itself. In the messy world of human affairs, things are always more complicated, so perfect self-regulation is too high a bar. However, threats to professional status might include incursions from patient autonomy movements, from governmental oversight, or legal regulation. If external sanctions are offered, it does not follow that something ceases to be a profession, but if enough external sanctions are offered—and they stick—it does raise a question about whether something is properly a profession. To illustrate this point, consider the difference between regulation that comes from within the profession, as opposed to that which comes from outside of the profession. Suppose, for example, a surgeon has a series of complications that lead to bad patient outcomes. Whether they are a host of common complications or a few rare complications, at some point other surgeons familiar with the situation may report the surgeon to an internal database or request that the surgeon be relieved of privileges at the hospital where they work. This would be an example of an internal sanction and could be understood as the profession regulating itself. In fact, if reporting surgeons cited one or two egregious mistakes with surgery-specific language, basing the level of mistake on the required surgical knowledge and skill, then the character of self-regulation would be well-illustrated. If, to the contrary, government officials who were charged with overseeing the care of populations noticed poor outcomes—e.g., the deaths of x number of patients at the hands of this surgeon—and cited, in revoking privileges to practice, the negative outcomes, this would be an example of external regulation. Enough of these kinds of encroachments, and the policies and additional justificatory work (from physicians) that might result highlight how external regulation—even if for good—can threaten the status of professionals.

The second conception (the sociological conception) is Buchanan's way of arguing that professions must justify their existence. He notes that professions often possess a "special status for members of the group and special privileges, including financial advantages and a significant sphere of autonomy, i.e., freedom from external regulation."[2] This can be read as a second conception because one might think that professions could exist without these benefits or this status; however, that many professions do possesses them makes it worth noting. Consider, for instance, special taxation structures or exemptions for institutions understood to be religious, or the special parking spots and significant compensation that (at least in some times and places) were offered to physicians. Buchanan's point here is that professions need to justify their reception of the benefits of this special status, if they possess such a status, because professions are not facts of nature; rather they are social constructs.[3]

With this, albeit incomplete, sketch of an account of professions, let us take up conscience, relative to the disagreement that can exist among professionals and, most importantly, how such disagreement might be resolved. But, first, the question of whether health professions, in particular the medical profession, exist must be addressed. The next section describes the founding oath of the medical profession and the connection between professions and practices (in the philosophical sense).

2.2 Do Healthcare Professions Exist?

If one accepts the five (or six) conditions laid out in the preceding section, and believes that each condition is satisfied then, yes, health professions exist. They are, as Buchanan notes, socially constructed. However, further argument may be beneficial for those who are suspicious of the motives of health professionals. This suspicion usually takes the form of a critique of the altruistic nature of the profession. If medical practitioners are not altruistic, then medicine fails to be a profession. Turning to the Hippocratic oath is useful here, in making the case for the altruistic nature of the profession. In so doing, the aim is not to claim that all physicians are altruistic, but rather that the focus on the care of and for others, makes it such that—in general—physicians must possess a dominant and intrinsic commitment to serving others; in

[2] Buchanan (2009, pp. 175–202).

[3] Buchanan (2009, pp. 175–202).

other words, the good physician—a member of the profession in good standing, we might say—exemplifies this trait.

2.2.1 *What Is Professed? Oaths*

The Hippocratic Oath, in its original form, is not commonly recited at medical school white coat or commencement ceremonies; if an oath is recited, it is a modernized version.[4] Yet, one might trace back from more modern recitations to the altruistic requirement at the crux of the original oath. Though the phrase most commonly associated with the Oath, "First, do no harm," is not in the original text.[5] The original Oath does state, "I will keep them [patients] from harm and injustice."[6] Ensuring that one's actions (or omissions) do not lead to harm can be understood as altruistic, but the original text makes the case more strongly, as the physician is not only supposed to guard against causing harm, but to keep patients from harm *and* injustice. The reference to patients reorients the obligations, emphasizing the object of the physician's care and, in so doing, fits even more deeply with a claim of altruism. Further, physicians, on this original reading, are also to keep patients from injustice. These obligations—depending on how one reads them—widen the scope of what is often thought of as medical practice (though current medical students and physicians who understand themselves to be advocates for their patients and place in their sights obstacles to realizing health, such as food desserts, climate change, and structural racism, are more in line with the original Oath than may be thought).

The aims of serving the patient exist whether one is examining the original Oath or in accounts of medicine embodied in current oaths that have been traced back to an account of medicine present in the original Oath. Allowing for differences in time, place, culture, and the development of medicine, one can find this powerful requirement to serve others there, which lives on in the more modern claims that are professed. One can read in the Oath's original prohibition on surgery, for example, the aforementioned point about injustice and harm, and recommendations with respect to diet as precursors to claims that the patient is a being who should not be harmed. In their oath, University of Chicago medical students declare, "May I never see in my patient anything but a fellow human in pain. My goal will be to help, or at least do no harm."[7] Thus this profession is one which marks, whether at a White Coat ceremony or again during commencement exercises, a dominant and intrinsic feature: keeping patients from harm and injustice is so strong that members of the

[4] For more on this subject, see Orr, Pang, Pellegrino, and Siegler's influential 1997 paper, "Use of the Hippocratic Oath: a review of twentieth century practice and a content analysis of oaths administered in medical schools in the U.S. and Canada in 1993" in the Journal of Clinical Ethics (8 (4) pp. 377–388). Students at some medical schools, for instance, recite the Physician's Pledge on these auspicious occasions.

[5] Edelstein (1943).

[6] Edelstein (1943).

[7] Searcy (2013).

profession strive to always see their patients as persons and vulnerable persons, at that. Also consider the Oath's requirement that physicians "come to the benefit of the sick" and its prohibition on engaging in sexual relations with patients. One might read this clause in a similar vein. There is something about physicians' patients, their fellows, such that harm is a violation.

This raises an important and timely question: What is harm? Here, the Oath is also helpful. One can read in the Oath an account of medicine as a practice[8] and as a profession. Though the primary aim of introducing the Oath at this point in the discussion is as a way of showing that altruism is baked into the very nature of the medical profession, the Oath does additional work in highlighting why medicine is a profession and offers potential resources for addressing disagreement within that profession. Prior to consideration of patients and after appeal to the divine, the Oath addresses physicians' relationship to other physicians, stating that information should be freely disseminated to only with those who have "signed the covenant" and "taken the oath."[9] These ideas serve as unifying features, which carve out a sphere for physicians and inform standards for practice.[10] In light of this, a physician might reasonably refuse a patient's demands for certain forms of treatment. Some treatment may harm "a fellow human in pain." Tracing this idea back to the Oath, one notices the prohibition on "deadly drugs" and "abortive remedies" as harmful interventions and, as such, not within the purview of the physician. Though this is not often understood to be the case today, practices of MAID and abortion remain contested spaces and, in some instances, practices about which physicians are allowed to step back from due to consideration of their conscience. Though the oath is in many ways out of step with current medical practice, a survey of the topics addressed in the Oath are less unfamiliar than might be anticipated by a first-time reader: prohibitions on assisted suicide and abortion, considerations of appropriate procedures, the fair dissemination of information, and concerns of privacy (medically relevant or not) live within the Oath and remain topics of discussion today. More to the point, it is neither the topics nor what a reading of the original oath might add to the ethical analysis of such topics that is essential to understanding the power of the Oath here; rather, it is that whether the details of the professed content is right or wrong, the profession of the oath is aimed at serving others. It is because, for example, that deadly drugs are understood to be harmful to the patient, that their prescription by physicians would be prohibited and not for another reason.

The Hippocratic Oath and its successors serve to unify the practice of medicine and to demarcate its purview: work in the service of the patient. Mention of this Oath at this juncture in the argument has another benefit, as well. It brings to the fore a set of topics about which current societal views have, broadly, changed, and so the

---

[8] It is worth noting that medicine is described in the Oath as an art. This notion is reemphasized in Louis Lasagna's 1964 modernized version of the oath (accessed at: http://guides.library.jhu.edu/c.php?g=202502&p=1335759). There is a great deal of work on the concept of practice; see especially MacIntyre (2007).

[9] Edelstein (1943).

[10] For a very nice treatment of these ideas, see John Brehany's (2011).

issue of disagreement becomes front and center. Physicians may (and do) disagree. Before turning to disagreement in practice, a topic I take up in Sect. 2.3, let us turn to a philosophical treatment of practices and their connection to the places where they occur—institutions—and the histories in which they exist—traditions—with the hope of bringing additional resources to bear on contemporary considerations of disagreement related to claims of conscience.

2.2.2 *Practices, Traditions, and Institutions*

Conceptualizing medicine as a practice that exists within a tradition may be helpful in understanding why disagreement exists within it.

Traditions are intriguing and often misunderstood entities. Any philosophical discussion of the concept of tradition would be remiss if it failed to engage the work of Alasdair MacIntyre, who has written two excellent books on the subject (*After Virtue: A Study in Moral Theory* and *Three Rival Versions of Moral Inquiry: Encyclopaedia, Genealogy, and Tradition*). MacIntyre's account weaves together the roles of traditions, practices, and institutions in an attempt to articulate a contemporary account of a virtuous life. For my purposes here, it is useful to think about medicine as a practice and useful for our discussion of medicine as a profession that it is that kind of activity, being enacted in institutions, and embedded within a tradition.

MacIntyre notes, famously or infamously, for those who study virtue ethics, that:

> By a 'practice' I am going to mean any coherent and complex form of socially established cooperative human activity through which goods internal to that form of activity are realized in the course of trying to achieve those standards of excellence which are appropriate to, and partially definitive of, that form of activity, with the result that human powers to achieve excellence, and human conceptions of the ends and goods involved, are systematically extended.[11]

There are important similarities between Buchanan's account of professions and MacIntyre's account of practices: both are socially established, both require excellence, et cetera. Further, though Buchanan does not stress this point, it lives within his work on the subject: both are complex! They are not isolated instances of medical advice or of the performance of a procedure. For Macintyre, chess is a practice, tic-tac-toe is not. Farming is a practice, haphazardly throwing turnip seeds in a field is not. Playing baseball is a practice, playing catch is not. Activities need to be broader, more organized, and more complex than the latter monad of the aforementioned dyads. Drawing blood is not medicine; reading a chart is not medicine; prescribing a medication is not medicine—though these all may be parts of the practical practice of medicine, they are insufficiently complex, organized, and broad to constitute the practice of medicine alone.

For MacIntyre, it is also important to distinguish practices from the institutions which house them. Here, though his work is not one in medical ethics, he uses an

---

[11] MacIntyre (2007). The definition of practice appears on page 187.

example of medicine; furthermore, understanding the relationship between practices and institutions, which are also not traditions, will prove useful:

> It is not only of course with sets of technical skills that practices ought to be contrasted. Practices must not be confused with institutions. Chess, physics and medicine are practices; chess clubs, laboratories, universities and hospitals are institutions. Institutions are characteristically and necessarily concerned with what I have called external goods. They are involved in acquiring money and other material goods; they are structured in terms of power and status, and they distribute money, power and status as rewards. Nor could they do otherwise if they are to sustain not only themselves, but also the practices of which they are the bearers. For no practices can survive for any length of time unsustained by institutions. Indeed so intimate is the relationship of practices to institutions – and consequently of the goods external to the goods internal to the practices in question that institutions and practices characteristically form a single causal order in which the ideals and the creativity of the practice are always vulnerable to the acquisitiveness of the institution, in which the cooperative care for common goods of the practice is always vulnerable to the competitiveness of the institution.[12]

We might, again, see a similarity between features of a profession and this philosophical account of practices. Especially relevant to the contrasting but intertwining nature of professions and the institutions that house them is MacIntyre's description of their different aims, kinds of goods, and tensions. These institutional features make practices vulnerable to those institutions. We can have concern for the creation of such vulnerabilities, while recognizing the necessity of those institutions for those very practices: one cannot, for example, operate without a surgical theater and all that comes with it, as well as the recovery rooms, scalpels, et cetera.

This brings us to the third helpful concept and the one which takes us the furthest in considering disagreement: traditions. A few points are essential for understanding traditions and how they connect to practices and institutions, but more importantly how they might relate to professions and disagreement. First, it is important to note that persons are the bearers of traditions, and so, given the focus of this text, physicians would be understood as bearers of the tradition of medical practice. MacIntyre writes, "What I am, therefore, is in key part what I inherit, a specific past that is present to some degree in my present. I find myself part of a history and that is generally to say, whether I like it or not, whether I recognize it or not, one of the bearers of a tradition."[13] Physicians cannot ignore the history of medical experimental abuses in interacting with patients; they cannot ignore its implications for trust or damages to trust, even if they would prefer to do so or even if they fail to recognize it.

For MacIntyre the subject of traditions is human reason. He writes, "For all reasoning takes place within the context of some traditional mode of thought, transcending through criticism and invention the limitations of what had hitherto been

[12] MacIntyre (2007, p. 194). MacIntyre continues this line of thought, writing "In this context the essential function of the virtues is clear. Without them, without justice, courage and truthfulness, practices could not resist the corrupting power of institutions" (AV 194). And so, we might think about the virtues needed for the practice of medicine and whether those virtues fit MacIntyre's analysis here and include the kinds of character traits that might allow one to stand up and advocate for one's patient—*keeping them from harm and injustice*—even if running up against the aims of the institution which houses the very practice of medicine that comes into question.

[13] MacIntyre (2007, p. 221).

reasoned in that tradition; this is as true of modern physics as of medieval logic."[14] Advances in medical science and of particular interest the increasing role that AI systems are playing in healthcare is a fitting example of the limits that were once thought to exist and have been surpassed, yet it is only in conversation with historical practices that one can understand advances from scalpel and suture, to the addition of cameras for laparoscopic surgery, of a third arm for robotic surgery, and for increased decisional capacity with algorithms augmenting human clinical decision-making.[15]

MacIntyre continues, "Moreover when a tradition is in good order it is always partially constituted by an argument about the goods the pursuit of which gives to that tradition its particular point and purpose."[16] That is, arguments and potential disagreements are not obstacles to practices, but a feature of the very traditions of which those practices are a part. Conflict can and will arise, and when it does, adjudication is needed. "So when an institution – a university, say, or a farm, or a hospital – is the bearer of a tradition of practice or practices, its common life will be partly, but in a centrally important way, constituted by a continuous argument as to what a university is and ought to be or what good farming is or what good medicine is. Traditions, when vital, embody continuities of conflict."[17] It is this conflict which leads to the refocusing of institutional missions, the reprioritizing of institutional goals, and attempts to understand again what it means to support a particular practice. Current investments in sustainable hospital infrastructure in the healthcare industry is one example.

Though conflict can lead to advancements and improvements, and potentially the kinds of excellences that are part of what it is to be a member of a profession, adjudication of or resolutions to particular conflicts are needed, and this is especially pressing in medicine. There must be an answer to a practical question, or the ramifications (e.g., possible harm to the patient) undermine the purpose of the profession in the first place. It is to this space, adjudicating disagreements within professions, to which we now turn. To frame those comments, this section concludes with MacIntyre's description of what a tradition, at least a healthy one, is:

> A living tradition then is an historically extended, socially embodied argument, and an argument precisely in part about the goods which constitute that tradition. Within a tradition the pursuit of goods extends through generations, sometimes through many generations. Hence the individual's search for his or her good is generally and characteristically conducted within a context defined by those traditions of which the individual's life is a part, and this is true both of those goods which are internal to practices and of the goods of a single life. Once again the narrative phenomenon of embedding is crucial: the history of a practice in our time is generally and characteristically embedded in and made intelligible in terms of the larger and longer history of the tradition through which the practice in its present form was conveyed to us; the history of each of our own lives is generally and characteristically

[14] MacIntyre (2007, p. 222).

[15] Pilkington and Binkley (2022, pp. 81–83).

[16] MacIntyre (2007, p. 222).

[17] MacIntyre (2007).

> embedded in and made intelligible in terms of the larger and longer histories of a number of traditions.[18]

Given this, how might professions both respect the consciences of their members, work through disagreement situated within long traditions of thought and practice, and also deliver (or contribute to the delivery of) excellent care? We can now turn to the topic of disagreement in earnest.

2.3 Disagreements in Practice: Adjudication, Reasons, and Reasonability

Claims of conscience are sometimes met with surprise as if they were unanticipated or seem out of place. That reaction itself is surprising to anyone familiar with human beings and with medical practice. Consider the diversity of medical professionals. The narratives of their lives vary. They train in different locations and institutions, the specialties they choose (and the cultures of those specialties) diverge, and the particular approaches to the health and care of patients that each adopts individually is distinct. This is not to say that there are no similarities nor to question standards of care; rather, it is to note that the existence of disagreement and for subsequent requests for respect ought not to be surprising. Further, any attention paid to recent trends in professionalism and the importance of professional identity formation in medical education will serve as one point of evidence for the importance placed on future physicians' construction of their own professional self-understandings in light of their own values and the demands of the practices of their profession. This is an important point of evidence as it arises from current practices within the profession itself and within an essential and influential area: the training of future members (which is even referred to, at least in healthcare professionalism contexts as, *formation*). That said, one might point to different stages in the medical profession's care of patients that would lead to the reaction of surprise when conscience is discussed or invoked. One might divide the interactions that constitute the care of patients into three stages. First, itinerant practitioners entered homes of persons to care for them, offer remedies for maladies, and address concerns associated with health.[19] In a second, much later stage, individual physicians treated patients as members of larger collaborations: engaging care in hospitals, not homes, and relying much more on the advances of medical science.[20] In a later stage, group practices arise, as physicians share the burden of care with others, and in a final stage, physicians are understood (or are on their way to being understood as) yet another member of a care team. To tell a history in a few sentences is dangerous as brevity can lead to mistaken interpretations and misunderstandings. Thankfully for our purposes here, what is needed for the

[18] MacIntyre (2007).

[19] Cavanaugh (2018).

[20] Al-Shura (2014).

argument to be successful is not a proper historical accounting of the progression of medical care, but rather the recognition that physicians have interacted with patients in different ways, in light of their involvement of other professionals of varying levels of sophistication, and with institutions. The solo (or apprenticed) practitioner traveling from one home to another may suggest a certain kind of ethic; for example, one in which confidentiality is extremely important. When in home care is paramount, keeping silent both about the health-related matters of the patient, but also about what a physician might come upon in the private spaces of those patients (their home, their families, their friends, and what occurs in those relationships and within those spaces), matters a great deal. This kind of confidentiality may not be prized (or even expected) by a patient who seeks care from an institution, as opposed to a person. Patient X sees a member of a care team, Dr. Y, who works a set schedule of hours and sees patients like X, not because X sought out Dr. Y, but because Dr. Y was assigned X. When asked about their recent visit to the doctor by their neighbor, W, Y shares with W that everything went well at *Z Health System*, neglecting to mention Dr. Y by name. These are two very different pictures of medical interactions and the ethical norms that ought to apply to each might very well diverge. It may be that conflation of earlier with more recent practices of medicine has led to the surprise or, at least, contributed to the disagreement which currently exists in medicine. In focusing on the former, it would be very surprising to not expect diversity; yet if one's experiences are primarily of the latter, it may very well be equally or more surprising to find requests that it be respected.

As argued in Chap. 1 of this book, disagreement in human life and especially about something so prized by so many—health—should not itself be surprising. However, the aim of the next set of sections is to examine not this reaction but the disagreement itself; in particular, it is to examine a popular (and in many ways, attractive) analysis of that disagreement and how it might be resolved within professions.

2.3.1 Reasonableness and Reflective Equilibrium

In this section, I focus on one prominent example of the conscience dissenter position and its approach to resolving disagreement. In an influential article in the *New England Journal of Medicine*, Ronit Stahl and Ezekiel Emanuel argue against protecting the consciences of healthcare professionals in most cases. They are interested in physicians, in particular, and believe that claims of conscience should be given no credence except for those that are in response to a professionally contested issue. Though this position does stand up to critical scrutiny, it is a valuable position to consider for at least three reasons. First, it captures a very common view on the subject. Second, the position is not an "all or nothing" position, which might allow for forward movement and potential resolutions within the debate over conscience; that is, it is possible that the class of exceptions is sufficiently large that the position allows for enough protection of conscience that a resolution between disagreeing parties becomes feasible. Third, because disagreement demarcates those instances in which conscience should not be appealed to and those in which it may be appealed to, there is an understanding of the diversity of persons and an attempt to give an account of and offer a process by which disagreement might be resolved. For ease, let us refer to this version of the conscience dissenter position as the Professional Association (PA) position. The reasons for this description will become clear in due course.

As proponents of PA, Stahl and Emanuel approach considerations of conscience as contractualists. They are, as many of us are, interested in the ethical practice of medicine, concerned with power differentials between physicians and patients, with informational asymmetries between those who provide care and those who are in need of it, and with the vulnerability of persons, especially those in a sickened state. Analyzing this situation in terms of a social contract is appealing, especially if one appreciates the history of professions, especially the medical profession,[21] and if professions (or, as will be argue later, professional associations) are the locus of focus and the onus of power in one's analysis of conscience.[22]

Proponents of PA believe that professions are the proper place for resolving disagreements around healthcare practice and that such a resolution about, say matters of conscience, are to be adjudicated there. Though proponents of PA reject most claims of conscience (for Stahl and Emanuel this is because, as noted in Chap. 1, claims of conscience derive their support from an analogy to military CO), they—fittingly—do allow for appeals to the conscience of a member of a profession in the instance that the profession has not determined what is appropriate, that is to say, in situations where the matter under consideration is contested. They write, with respect to "interventions about which the health care community is debating whether participation is appropriate or not," appeals to conscience are permissible. On this view, professional standards are what guide (or ought to guide) physicians, not a physician's conscience, and so it makes sense that if there are professionally contested interventions, the PA position would have to leave determinations about these up to something or someone else. There are no normative resources within this account to do otherwise. Were they to bring in other theoretical resources to assess the status of new or contested interventions, then those resources could be applied to other interventions, as well. However, they claim that all other matters that have been "settled" are not up for discussion. Rooting the normative analysis in professions thus cuts off other possibilities. The PA position roots this normativity in professions because, according its proponents, professions, not individual professionals, determine what is in the primary interest of the profession. The primary interest of the medical profession is "to promote the well-being of patients."[23] The

[21] Starr (2017).

[22] A contractualist approach fits this analysis of power dynamics, as well, and the concern for the vulnerability of patients. In examining the relationship between physicians and patients, a contractualist might describe these sets of concerns as highlighting a need for parties in an agreement to be equally situated. It is only when parties are on a similar footing that they can properly enter into a contract and that the contract can be properly enforced (in this case, that it would be binding on all parties).

[23] Stahl and Emanuel (2017, p. 1382).

interpretation of what it is to promote and what conception of well-being is operative are to be determined by the profession.

Given the arguments in the earlier sections of this text, it should be obvious why this is an attractive possibility. Professions are taken very seriously, at least by some persons. However, this position is referred to as the Professional Association position, not the Profession position, because in attempts to adjudicate potential disagreements, it becomes clear that proponents of PA are not thinking of professions but rather of professional associations. It would not be the profession of medicine but a professional association, like American Medical Association (AMA) in the United States or, one could imagine, the World Medical Association (WMA) for global considerations, which would make these determinations and wherein debate about interventions and practices would occur. If one thinks of medical specialties (or subspecialties) as sub-professions,[24] then it would not be the surgical profession, but the American College of Surgeons (for issues arising in the United States) who would determine what is in its primary interest and it would be the discussions taking place there by members (or their representatives) that would determine interventions that might remain contested and about which appeals to conscience relative to such interventions would be admissible.

One of the many benefits of a contractualist approach to issues related to conscience is that it restructures what might first appear as a variety of disagreements at a number of different levels among diverse persons to a single set of disagreements: claims arising from a personal sphere and those arising from a professional sphere.[25] In particular, proponents of PA (and this is very much the case for Stahl and Emanuel) are concerned with the introduction of personal philosophic—but most especially of religious—considerations into the deliberations of a professional association and, derivative of that association's primary interest, the well-being of patients. These arguments are taken up in the next chapter, but it is important here to lay some of the contractualist groundwork for why this division is accepted (if not argued for) first. In laying this groundwork, two points must be emphasized: first, what is the operative conception of professions in this account—is it something akin to what was described in the preceding sections or is it something distinct? Second, what is the root of disagreement? What is the process of alleviation of that disagreement that proponents of PA offer? Is it successful? Let us consider each point in turn.

The way in which proponents of PA understand professions appears to depend upon a less sophisticated view of disagreement or the topics of disagreement within the profession of medicine than is needed. Current medical practice is deeply informed and guided by standards of care, but to practice in line with standards of care does not mean that one cannot appreciate different approaches to medical challenges, different positions, distinct influences, varying styles of interactions with

[24] This notion of a sub-profession should be understood loosely. The point does not hang on whether or not one endorses the plethora of more particularized "professions" that might exist if there were things like sub-professions. Rather, the invocation of sub-professions is for the purpose of highlighting that it is the members of these types of societies and not any member of a profession, which is a person who practice the craft associated with the profession, which counts.

[25] Consider, for example, Rawls' work on public reason (see especially Rawls [2005]).

patients, et cetera, nor does it mean that standards of care are indefinitely fixed.[26] Additionally, as noted above, the PA position references professions but – and here is the first point of the two – it, in fact, appears to have targeted professional associations, not professions as the relevant arena of debate. To rely on a description of "the medical profession" but to use professional associations as conceptions thereof that do the argumentative heavy lifting[27] is to involve an equivocation. It is an equivocation whereby proponents of PA mean professional societies or associations but act as if they are referencing professions. In fact, though Stahl and Emanuel discuss the medical profession in a broad manner, the evidence that they offer and the examples that they appeal to in their analysis are rooted in things like the codes of, and guidance given by, professional associations. For example, they accuse the AMA of being "internally inconsistent regarding conscientious objection."[28] They also find fault with the American Pharmacists Association, describing their codes and guidance as "ambivalent."[29] Further to the point, in addressing issues relevant to conscience, they consider the American College of Obstetricians and Gynecologists and the American Psychological Association, which are additional examples of professional associations, not of professions themselves. They proceed from consideration of these examples to considerations of health care providers and "the profession,"[30] but such associations are not the same as nor are they even coextensive with a profession (or even sub-professions). This is problematic; it shows that the PA position, at least this articulation of it, rests on a faulty assumption. Thus, undergirding the entire attempt to resolve disagreement and to address the issues related to appeals to conscience in medicine is a mistaken and, ultimately, fallacious claim. Now, let us consider the process of adjudicating disagreement before returning to a discussion of professions and professional associations.

According to PA, if disagreement arises, such conflicts can be resolved through a process known as reflective equilibrium. The process of RE is a common one in political theory and there are numerous articulations of it. At its core, RE is a response to cognitive dissonance, of a sort. It encourages those who might avail themselves of its benefits: to examine particular principles that are held, then to examine particular cases that arise, returning to those examined principles that might bear on the cases in questions, with the goal of having one's judgements of cases and the principles that one holds cohere (or to fit together well enough as to allow one to offer consistent judgements).[31] Thus, if one employs this process, one might be able to bring together judgements about challenging cases through comparisons to other cases by means of principles. For example, suppose a medical professional in training holds the view

[26] Any person familiar with progress in medical sciences and the advances due to scientific discovery and innovation sees first-hand the great achievements of research in the medical sciences and the advances in standards of care due to that research.

[27] For an earlier discussion of this point, see: Pilkington (2021, pp. 233–245).

[28] Stahl and Emanuel (2017, p. 1383).

[29] Stahl and Emanuel (2017, p. 1384).

[30] Stahl and Emanuel (2017, p. 1384).

[31] A nice treatment of the idea can be found in Huemer (2008).

that medical professionals should always help the greatest number of persons that they can (within the limits of time or a proper work-life balance or something similar). As the trainee progresses moving from a medical student to becoming a physician, they begin to understand better the complex systems of healthcare delivery and how they are bound up with policy determinations, resource acquisition and allocation, as well as non-medical determinants of health. This trainee may come to find that they may be able to help more persons in need by supporting or engaging in research which harms a set of vulnerable persons.[32] By availing themselves of the process of RE—moving back and forth between considered judgments of cases and the principles that they espouse, the now physician need not jettison their principles (nor fall back to some kind of nihilistic or, worse, inconsistent position) but rather revise them in such a way that they cohere with their considered judgements (or vice versa). Thus, a principle like "Always help the most patients." may become "Help the most patients without harming any patients directly." This is a fairly simplistic example, but it captures the general idea of RE. It is important for the argument in this section because contractualist approaches often appeal to something like it and PA relies upon one particular version of it. With this tool at the ready, let us turn to the disagreement.

PA's explanation of the cause of disagreement (and this is the case for many conscience dissenter positions[33]) is that some members of the medical profession bring to the discussion of the norms of their profession their own personal, personal philosophical, or religious views. They then attempt to offer these views, either directly or as the basis for their claims, and that in doing so they offer norms that come into conflict with the previously stated or determined norms of the profession. Disagreement in this instance can be resolved, and Stahl and Emanuel appeal to John Rawls' version of RE. They write, "This dynamic process establishes professional obligations for health care providers regardless of their personal beliefs."[34] Thus, or so they argue, protecting the conscience of persons who would offer these kinds of reasons, results in bad medical practice. This is because, as they argue, "In a professional context, personal religious convictions are secondary."[35] Doing work behind the scenes of this claim is the aforementioned distinction between what is to be held personally and what can influence or be offered to others. In Rawlsian language, this is the distinction between private and public reasons.[36] Stahl and Emanuel's claim is illustrated with a set of cases; they write, "… the Jehovah's Witness surgeon cannot refuse to allow blood transfusions during surgery. The Jewish pharmacist cannot withhold pills that are made with nonkosher gelatin. The Mormon nurse cannot refuse to treat alcoholics."[37] Their assessment is understandable, especially

[32] The history of research and research ethics is replete with examples. For a recent treatment of such cases, see: Elliott (2024).

[33] For one place in which these ideas are taken up, see: Pilkington (2021, pp. 233–245).

[34] Stahl and Emanuel (2017, p. 1382).

[35] Stahl and Emanuel (2017, p. 1382).

[36] Rawls (2005).

[37] Stahl and Emanuel (2017, p. 1382).

given the tool that is to be employed. If RE is to be effective, the movement must be between relatively similar principles (or, at least, principles which could be revised in such a way as to be relatively similar). The process would not work were that not the case. Further, since they understand the profession—though they actually rely on professional associations—as the unit in which the process of RE would occur, the fact that the profession has determined certain standards, it follows that all members of the profession must abide by those standards.[38] Thus, particular features of individual professionals, such as their religious commitments or their personal philosophical views are not admissible, and must be divorced from their professional identity or, as Stahl and Emanuel note, physicians must leave their religion[39] at the door.

This analysis, and ultimately the PA position, fails and does so for a number of reasons. First, and most germane, is that the process of RE works well for a person seeking to resolve tensions between ethical assessments of challenging cases. It is not clear that it works well as a way for a group of people to resolve disagreements. In order to adjudicate disagreement by means of this process, an argument is needed for why one set of considered judgements and principles, as opposed to another set of considered judgments and principles, must be chosen. That is, the process does not operate in a vacuum. This consideration is obscured if the aim of the process is to relieve cognitive dissonance in a single person or to harmonize discordant or in-tension features in a single unit. However, individual professionals comprise professions (or professional associations), and they hold complex views and live rich moral lives.

To put a finer point on this argument, consider concrete cases about which members of a profession(al association) might argue. Stahl and Emanuel note that when cases or issues are contested, then physicians may conscientiously object. However, what if this is a sizable number of cases? Take, for instance, disagreement about the ethical status of cardiac devices: is it permissible to turn off such devices when they are properly functioning and the patient is "pacemaker dependent," if so, may a physician who does not wish to tun off such a device be allowed to practice conscientiously (which, in this case, would mean refusing to turn off the cardiac device in question)? This is a potentially helpful example, as Stahl and Emanuel offer MAID as one of the issues that is not yet settled and at the heart of the disagreement over the deactivation of some devices is the question of whether patients or their clinicians are involved in their death or in hastening it. If turning off a device constitutes a patient taking her own life, then should physicians be permitted to assist? Barred from assisting? Allowed to not assist? Whose considered judgments are the

[38] If one accepts my critique that PA is really about professional associations and not about professions, then members of the professional association are those who would be bound by its dictates. That these dictates are only binding on some relevant professionals suggests that the argument for PA is not terribly strong.

[39] Stahl and Emanuel are not the only CDs to hone in on religion. The abstract of Savulescu's influential article, "Conscientious Objection in Medicine," reads as follows: "Deeply held religious beliefs may conflict with some aspects of medical practice. But doctors cannot make moral judgments on behalf of patients" (Savulescu 2006, pp. 294–297).

ones which should carry the day? On this particular topic, should all physicians be surveyed, and their considerations be input into the RE process? Should physicians who focus on cardiac issues that are often treated with devices be given special priority or should their judgements be more heavily weighted? Physicians do not operate alone and given that much of healthcare is team care, should the considered judgments of nurses whose work focuses on cardiology be considered? Should the industry employed allied professionals (that is, the "techs," who, according to the literature do more deactivations than either of these other groups, have their voices heard?[40,41] Depending on one's answers to these questions, one might seek assistance (were one working through these ideas in the US) from the AMA, the Heart Rhythm Society of America, or the American Nurses Association. But these are all different professional societies with different memberships, distinct views, and profession-specific concerns. The point here is not to adjudicate this disagreement, but to suggest that the process of RE might not be sufficient to address the complexity of the landscape, and this is only one kind of case, which happens to touch on matters that are deemed to be contested (assisting in death).[42]

This kind of moral diversity does not appear to be a great fit for a fix by the tool of RE. This is especially true because in addition to these disagreements the expressed views and supporting positions are informed by distinct traditions of thought, various lived experiences, and other professional commitments. However, it may be that a general discussion of these points is insufficiently persuasive and what is needed is a deeper dive into the kind of RE that PA relies upon.

Rawlsian RE is referenced by PA as the means by which disagreement might be alleviated. The concern with its employment is not that the process cannot be useful, though questions can be raised about its applicability in this context; rather, the present critique is that proponents of PA have neither shown why this is the best process nor who should participate in it. This is a weighty matter because, as argued above, it is not merely the process through which a single person moves back and forth between established principles and their evaluation of particular cases, but one where a (potentially very large) group must do so. Does the Rawlsian emphasis add sufficient depth and additional resources to overcome these concerns? The answer is no, and a more careful look into Rawls' work suggests why.

For Rawls, parties to agreements are to be fairly situated, reflective of the fact that such parties are construed by Rawls as both free and equal.[43] The conception of freedom that Rawls offers stresses three components. First, persons must possess

---

[40] Mueller et al. (2008, pp. 560–568).

[41] Mueller et al. (2011, pp. 253–261).

[42] The actual ethical and professionalism-related ground is even more complicated. Some professionals distinguish between certain devices or certain functions (for example, distinguishing pacemakers from defibrillators), while others do not. Some professionals rely on philosophical distinctions between withdrawing and withholding, but others do not. Some professionals focus on the intention of the agent, while others view such a focus to be insufficient, and still others disagree about what the content of the intention is. For a full discussion of this issue, see Pilkington (2020, pp. 28–41).

[43] See Rawls (1999, 2005).

their own conception of the good. Second, persons must take responsibility for their ends. Third, persons are to be understood as self-authenticating (or, sometimes, self-originating) sources of valid claims.[44] For those familiar with Rawls' work, which is greatly influential in political philosophy, legal philosophy, and ethics, this is merely rehearsing key points from his views on political liberalism. However, those less familiar with this work but familiar with conscience dissenter positions, like PA, might find these features to be surprising. To reject protections for conscientious practice but offer a process to adjudicate disagreement rooted in Rawls' work is an odd argumentative move. In order to support this concern, let us consider each of these three components to see how proponents of PA incorporate or address them. If they do not, then it is unclear why Rawlsian RE is the right fit for the account and, further, an appeal to it does not appear to be justified. Before engaging these notions, one possible response should be addressed. Though this argumentative move is not made by Stahl and Emanuel, a proponent of PA might argue that PA can, and does do this work, but that it is the freedom of patients, their conceptions of the good, and their obligations to take responsibility for their own ends that support the use of Rawlsian RE. Though it may be true that a CD, even a proponent of PA, aims to do this, it is not relevant at this juncture of the argument. Though this approach makes sense, given the focus on the well-being of patients, what is needed to move the argument forward is an application to medical professionals, not patients. This is not because the well-being of patient is not of significant concern; it is, after all, the object of the profession, but rather because PA has placed the adjudication of disagreement within that very place, the profession (though we have understood that by that claim, the proponent of PA must mean something "within a professional association or society"). And so, the process of RE must take place between the members of that profession (or that professional association), even if patient well-being is understood to be the aim of the profession. Patient perspectives, however important, are not the perspectives in tension. The disagreement that is the object of attempted resolution is among professionals. With that potential objection avoided, let us now turn to those three Rawlsian components of freedom to see if PA can do them justice and thus employ a version of Rawlsian RE.

To claim that persons must possess a conception of the good in order to be free is an ill-fitting claim for the conscience dissenter position as—in a variety of articulations, but especially according to PA—medical professionals are not understood to possess this kind of conception within their professional context. It is true that one might espouse PA and separately hold to a generally Rawlsian political philosophy, but the latter is being employed within the former at a crucial step in the argument for PA. The second component, that persons must take responsibility for the ends, can be interpreted in a number of ways within this context and, on some of those interpretations, the conscience dissenter position could espouse a Rawlsian view. If one interprets this claim in the context of debates over conscience in medicine, one might see responsibility for ends leading to requirements that those who wish to practice conscientiously disclose details of their practice in advance, that they

[44] Rawls (2005).

offer reasons for their conscientious practice, and the like.[45] These requirements fit well with position in between the poles, as well as with conscience protector positions. They do not fit as well with conscience dissenter positions, however, as those positions do not leave room for professionals to take responsibilities for their own ends—rather they are to follow the agreed upon view of the profession (read: professional association). That said, one might interpret this Rawlsian component as simply needing to offer arguments in those debates that might occur over contested interventions; that is, taking responsibility for this truncated set of professional ends. If that is the case, then the proponent of PA might have a better claim to a Rawlsian label.[46] However, it is challenging to see how the third component could be adopted by PA advocates. This is because to do so would require that they think of persons self-authenticating or self-originating sources of valid claims and to do that is not how CDs conceive of professionals. Professionals lack exactly this feature within professional contexts. Their claims are rooted in something like the *expression of their professionality*. It should be noted that some CDs frame the tension as one between physicians and patients, and with that framing, one might think of both sets of persons as needing to treat the other as a self-originating source of valid claims, and then seek a way to choose which set of claims trumps the other.[47] Some of these positions are addressed in other places in this text. However, this move is not available to the advocate of PA, even if it might be available to a non-PA CD, as it is disagreement among professionals that needs to be adjudicated and not disagreement between patients and medical professionals. Still, a CD might push back, arguing that members of professions have opportunities to engage in processes within their own professional association. Additionally, another CD might argue that professionals, like all citizens, have a say in societal laws by means of their vote. However, these are not sufficiently persuasive replies to rebuff the critique. Putting to the side the need for further clarity about the aforementioned equivocation between "the profession" and "professional associations," for this reply to be sufficient, much more is required than a brief discussion of RE. The approach taken by proponents of PA, at least in this version of the argument, is both dismissive of those who wish to practice conscientiously and may even beg the question against the CP. It may beg the question by returning the discussion to the professional association in order to employ RE as the adjudicative mechanism, but it is exactly this issue that is under scrutiny.

A resourceful CD might prefer the second reply: vote! This reply relies on engagement with the political process as the appropriate means for regulating a robust and

[45] These and requirements like it are sometimes known as publicity conditions.

[46] However, it should be noted that this is not the strongest interpretation, but a fairly liberal one, offered in order to try to make the case for the best fit, yielding the strongest PA position.

[47] This description fits the work of Greenblum and Kasperbaur, see: Greenblum and Kasperbauer (2018, pp. 313–317).

nuanced practice that forms part of a person's identity, not merely their professional identity.[48] A better explanation for what is occurring is that there are different conceptions of medicine operative in the analysis offered by the CD and what many CPs would understand medicine to be. In the background of a conscience dissenter position, like PA, may be an account of the practice of medicine wherein medicine is treated as or understood to be a commodity or a service. If this is the case, an argument—and a strong one—is needed. There might be good reason to think that medicine is, should be, or has now become something like a commodity. But, if that is true, then the professionals in question are no longer members of a profession. Terms like "physician," or "nurse," no longer refer to a vocation; they are merely occupations. Physicians, like bricklayers, offer something for sale, and patients—now to be known as consumers—can buy it. The old adage that "the consumer is always right" is a fitting description of how some CDs appear to understand the physician–patient relationship.

Thus, for these reasons, PA cannot hold up Rawlsian RE as its means to adjudicate disagreement and, as a result, PA fails. It fails because it needs an account of adjudication of disagreement within professions, as it allows for issues that are contested to be the (and the only) interventions about which conscience may be appealed. I now turn to a discussion of professional associations and highlight differences between them and professions, ultimately adding to the argument that they are not the same thing.

2.4 Professional Associations Are not Professions

Arguments throughout this chapter rely on the idea that professional associations are not professions. Given that the philosophical heaving lifting in the arguments of many CDs is done by a conception of professions (or, in instances where no conception is articulated, an implicit argument of the form: because physicians are members of a professions, and professional membership requires *x*, then physicians ought to *x*—wherein x precludes the possibility of practicing conscientiously), this is a crucial topic to address. Sometimes this argumentative structure is populated with content about the profession. For example, on some accounts, members of professions are described as public or semi-public employees.[49] (These accounts are addressed later in the text.) This section focuses on the connections between professions and professional associations. This is a key step in the overall argument because if it is the case that these two kinds of entities are similar enough, then PA would not fall to the objection that it equivocates on the notion of profession. As has already

[48] On the importance of professional identity and the formation thereof in academic medicine and medical schools, see: Cruess et al. (2014, pp. 1446–1451).

[49] See: Greenblum and Kasperbauer (2018, pp. 313–317) for one example of this contention. See also: Savulescu (2006, pp. 294–297) for an earlier and more robust articulation with similar consequences.

been illustrated, PA relies on professional associations, and not on professions, in its analysis. Are they similar enough for that argument to work? The answer is no.

Consider the diversity of professional associations and societies associated with the medical profession (if one considers the number of professional associations and societies associated with all healthcare practice, the list is even more numerous). Medical professionals comprise quite a significant list of professional associations, given the kinds of practice—whether they be traditional, integrative, holistic, et cetera; whether they focus on a particular population, say older persons (geriatric medical societies) or younger persons (pediatric medical societies), whether they address issues connected to a particular region (the New York association of medical *x* or the San Francisco medical society of *y*, or wider region/medicine combinations, such as those societies focusing on "tropical medicine"). The point here is that there exists a great deal of medical societies, but there is said to be a single medical profession. Even if one tracks the history of different groups vying for the socially prized place of offering care to others,[50] and so might be tempted to claim that there exists DOs and MDs, trends are such that these are, in many contexts if not completely, interchangeable currently). How is the proponent of PA to reconcile the numerous societies with the claims in their argument about the profession? Should PA root its analysis of conscience in medicine and its attempt to adjudicate disagreement across all medical professionals in the work of, say, the American Osteopathic Association, or the American College of Obstetricians and Gynecologists? What of the Heart Rhythm Society or the American Academy of Neurology? Should the New York Academy of Medicine or the American Academy of Pediatrics—and this is to mention only a few societies housed within the US. Were one to look across the globe the list would be significantly longer. This list is also only populated by plausible candidates for PA and does not include, given the claims of PA, associations which might argue that they themselves should have some say in these conversations or that their members should have standing or possess a status that ought to allow them to join deliberations about conscience in medicine and interventions which might be contested. Consider, for example, the Islamic Association of North America or the Christian Medical and Dental Association; what of the American Jewish Medical Association or the Catholic Medical Association?

The proponent of PA might reject this list for a variety of reasons, but the strongest would be that any of these organizations lacks the breadth to be reasonably thought of as a stand in for the profession of medicine. If we remain in the US context,[51] the AMA is the most likely candidate to aid the PA argument. The AMA, as the professional association applicable to all physicians in the US, would presumably be that society that comes closest to being coextensive with the profession.

---

[50] See Starr (2017).

[51] This scope restriction exists for ease, given the focus of this project. Were this not the case, an organization like the WMA would be an excellent candidate for the PA position. Interestingly, the WMA was considering, discussing, and debating the role of conscience in medicine in 2022 and the topic was highlighted as one of, if not the, key issues in medicine. (Meeting of the International Chair in Bioethics, Porto, Portugal, 2022.).

Not only is the AMA the most fitting candidate for position of a professional association that might do the necessary work to count as a profession, but it is also the first association to be invoked by Stahl and Emanuel and thus might be exactly what that set of proponents has in mind when thinking about professions. Let us examine the AMA as a medical profession. Though it is the strongest option, it too fails to satisfy the needs of PA. The primary issue is that it is not the proper arena in which disagreement of the type that arises in the practices of medicine ought to be addressed. It is also not the right place to seek guidance on the kinds of cases which are envisioned by the proponent of PA, even if it is the right place for guidance of other kinds related to medicine. This is for three reasons:

First, in order for a single professional association to serve as (or in the place of) a profession, it would have to be able to perform functions of a profession; that is, it might have to reasonably and reliably speak for the members of the profession, offer relevant guidelines and advice regarding a variety of topics that impact the professional (that is, profession-specific) practice of medicine, and—most importantly for our purposes here—it must be able to serve as a means of adjudicating disagreement among professionals (or provide the relevant resources to do so). In order for a professional society to perform these tasks it stands to reason that professionals identify with the professional association in a way that is similar to how they identify with their profession. A reasonable indication of such identification is membership, that is, if the proponent of PA is correct, then *at least* a majority of physicians should identify with, that is, be a member of the AMA. However, most physicians do not belong to the AMA.[52] In light of this lack of membership, it is unclear how if the AMA—the reasonably best candidate for the job—does not work, another would.

The first reason is decisive; however, there might be conceptions of identification that are prioritized or privileged in such a way that membership in the organization is not a proper placeholder for that kind of identification.[53] For clarity, let us consider the subsequent two reasons. The second reason that the AMA fails as a candidate is that such a candidate must be the proper location for the adjudication of disagreements among professionals, and for the PA position this means that it is the right place to employ the process of RE. As argued above, contested interventions (e.g., those involving assisted dying) are often rooted in concrete cases and just as professional associations, generally construed, failed to be a fitting venue, so too will a particular one, such as the AMA. Locating this disagreement between members of the AMA will not do anything differently to resolve the disagreements or to make the process better able to answer the aforementioned questions, directed above to a professional association, generally construed.

The third, and final reason, that the AMA fails as a candidate is that even here, in an organization with only one quarter of the physician population as members,

---

[52] Graham, Judith. "'Like a Slap in the Face': Doctors No Longer Feel the Nation's Largest Doctors Group Represents Their Interests." Business Insider. Accessed July 18, 2024. https://www.businessinsider.com/doctors-american-medical-association-2016-12.

[53] It is worth noting that it is not the case that one must be a member of the AMA to practice medicine in the US, and so, it is not as though concerns about coercion would prevent membership from connecting to identification.

physicians are still persons. They still bring with them, even after medical training (including both its explicit and its hidden curricula[54]) and other forms of professional education and socialization, their own judgements about concrete cases; those judgements are rooted in various reasons and rationales, of various sorts, emerging from diverse traditions of thought, and lead to different (even distinct) normative implications. This places a challenge on the CD, in particular the proponents of PA, to explain how such a diversity of reasons might be worked into a Rawlsian schema. This is not to say that it cannot be done. There may be Rawlsian resources available to the PA proponent—restrictions on comprehensive doctrines, arguments in favor of the exclusive use of some variant of public reason—maybe *professional reason*, by which it might be meant everything that public reason entails but where the public is constituted only by the profession? However, even here, it should be noted that, given membership numbers, it may be *professional association reason*, not the reason of professionals.

Thus, if the PA position is to succeed, it needs a response to the first concern, about the association, as well as a response to the second (a concern about the membership of the association) and to the third (how to make sense of and properly apply the resources of the entire Rawlsian system. In acting as if a professional association can stand in for the profession of medicine, the PA position falls to an equivocation or, at least, assumes a much greater degree of homogeneity within the medical profession than exists in actuality.

2.5 Burdens of Proof

This final section this chapter serves both as a culminating point of the chapter and a transition to Chap. 3. As noted above, conscience dissenter positions of various stripes rule out reasonable accounts of the moral life—that is, they fail to take seriously the fact that persons, including physicians, possess rich (both nuanced and complicated) moral lives. Some of these issues associated with this claim are articulated earlier in this chapter and also in Chap. 1. Later on, it is argued that conceptions of moral responsibility fit into this category. The focus of this section, however, is the notion of integrity as a means for understanding what might be ruled in and out—via PA's preferred philosophical approach, a Rawlsian one—ultimately arguing that the burden of proof lies with the CDs and not with the CPs or the person appealing to conscience in this debate.

To illustrate these points, consider the line from proponents of PA that medical professionals may not rely on "personal" or "personal, religious" reasons as the basis for their professional practice or their adjudication of disagreements within the profession. Let us put this line into conversation with a very popular defense of conscientious practice: that it allows for the retention or maintenance of the moral

---

[54] The former is relatively straightforward. For more on the latter, see: Hafferty and Franks (1994, pp. 861–871).

integrity of medical professionals.[55] Per Wicclair's influential arguments, the tension that arises exists if there is a conflict between one's "core beliefs" and beliefs that undergird the practices of the profession.[56] If preserving one's moral integrity is a good, then there is reason to believe that space should be afforded medical professionals to (or at least attempt to) do so; however, if it is not a good, then little reason exists to carve out such a space for the others to bear the costs of some professionals living in such a space. Consider an example.

A physician claims that they are not capable of participating in a particular intervention (and suppose it is not on PA's list of "contested interventions") because to do so, they believe, would violate their core beliefs, and, in so doing, they would risk rupturing their moral integrity. Further, suppose that this medical professional offers an explanation for this decision that is rooted in a faith tradition (and that the content of those reasons is, at least in part, religiously oriented). This type of case is a common one in discussions of conscience in medicine. The aim of reminding the reader of this type of case is not to begin to settle the relevant issues, that would take another book (one focused on reasonable accounts of the moral life); it is to illustrate again that the conceptual resources that the CD (at least in the form of the proponent of PA) employs are insufficient to adjudicate disagreement. CDs, in general, and proponents of PA, in particular, reject reasons rooted in persons (not in the profession, legal system, et cetera) as admissible into the process of RE. The CD would be correct in asserting that, in the case under consideration, the reasons attributed by the medical professional would not be admissible on a Rawlsian account, as they would be part of a comprehensive doctrine. As this is not a text on interpreting Rawls, we might think of the distinction between public and private reasons in the following, albeit simplified, manner: we can conceptualize particular personal religious and philosophical views as parts of a person's comprehensive doctrine. Further, we might differentiate reasons rooted in these doctrines, with public, or more broadly applicable, reasons. We might say something similar of rationales or more full explanations of reasons and why they are held. Consider, for example, a common case to illustrate the difference. Suppose a citizen offers an argument in a town meeting in support of a prohibition of businesses being open on Sunday in that town (or on Saturday or another day). Suppose further that the reasons offered for this are that (a) Businesses ought to be closed on the Sabbath because the Bible specifies that work should not be done on the Lord's day. Suppose another citizen offers an argument for the same proposal but does so for reason (b): research in medical science suggests that days away from work are part of a healthy lifestyle and so business should be closed to citizens one day a week, and Sunday is a day of the week. (a) would not be admissible on a Rawlsian account, but (b) would be admissible.[57] What is interesting for our purposes here

---

55 Wicclair (2011).

56 Though this is a popular and well-defended argument, it is not agreed to by all interlocutors in the debate. For a critique of this position, see Byrnes (2021, pp.291–305). For Wicclair's response to this argument, see: Wicclair (2021, pp. 307–324).

57 For a full discussion of comprehensive doctrines and other relevant Rawlsian resources, see Rawls (2005).

is a third option. Suppose in a discussion of the rules of the town, another citizen rejected reasons like (a), noting that these were personal reasons and not the kinds of explanations which would be agreed upon or even understood by all members of the town. However, they noted that many members of the town prize their own religious views and practices and so suggest that the town consider carving out some kind of space for religious practices or make allowances for religious views to be considered, in general. Call this latter claim: (c). (c) *is* admissible on a Rawlsian account. (c) can be understood as a public reason and, it is worth noting, it has no religious content. The townsperson who advocates for (c) could very well be using sociological data or simply have been anecdotally informed that other townspersons care about religion. For our purposes here, it is worth distinguishing (c) from (a) and (b). This is not because it is a different kind of claim, being about those who care about religion, not about personal religious practices themselves, but because this kind of claim would be admissible on a Rawlsian view but not on the CD's (at least not on the PA proponent's) view. Thus, there is another issue with PA, in that it again seems to come apart from its self-claimed philosophical moorings, at least with respect to disagreement.

It may be worth underscoring the points of the preceding paragraphs. Though personal views, and in this case personal, religious views would not be allowed into the discussion of contested interventions at the professional association, they should be allowed on a Rawlsian framework. A good Rawlsian, like the townsperson who advocated for (a) will recognize that many persons hold religious views, practices, and traditions to be of significant value. This is not to claim that a good Rawlsian would hold that reasons with religious content could be offered on the floor of the professional association's meeting. Rather, a good Rawlsian would admit reasons (and their accompanying arguments) which acknowledged the value of religious reasons in the lives of members of the professional association, as persons. Arguments about capping the number of hours in a work week so as to allow for religious or personal philosophical engagement and practice or arguments about the inclusion of physical spaces for such practices that already exist in hospitals (consider, for example, hospital chapels or interfaith rooms) would be admissible. The CD, insofar as they rely on a process of RE that is Rawlsian, has not opened the door to religious claims grounding medical practice. However, they have opened the door—and should hold it open, especially without offering an argument to the contrary—to the recognition that personal philosophical (including religious) reasons are of great significance to numerous persons (and that number includes physicians), impact the development of their self-conceptions, and may even contribute to the reasons that they form their professional identity in particular ways. Thus, this classification of personal views (including religious views)—not private reasons but public reasons which engage personal and religious views—should be admitted into the discussion.

If the argument of the last few paragraphs is sound, then why should a Rawlsian and proponents of position that invokes Rawls' RE be in favor of the moral integrity defense? Let us let this idea play out, at least in the form of a thought experiment, and see where it leads. If members of a profession (that is, professional association) were to take part in RE, reliant on Rawlsian resources, they might settle on a principle or considered judgment that safeguarded a practitioner's moral integrity. The CD, in PA and in other articulations, does not. What would they say in response to this departure? A few options exist. First, they might simply abandon RE as an adjudicative process, or they might abandon the resources offered by Rawls' political theory in total. They may, and rightly, argue that my line of critique, steeped in Rawlsian interpretation (even if at a basic level) is too philosophically laden[58] and that depth of analysis is not needed for a feature of their argument that is actually external to the conclusion at which they are interested in arriving. That is to say, in the way in which one might describe the failure of their chosen adjudication process to be an *internal* problem for their account. Ruling out moral integrity and disregarding personal views is not internally problematic, that is true, as it would not raise a problem of coherence as an issue that is internal to that account would. Such a determination would make the conscientious dissenter's position much, much less attractive. This is the case for at least two reasons. First, it leaves the proponent of PA without a means by which to adjudicate disagreement, and some mechanism for doing so is sorely needed. Given how the account is constructed, including the recognition of disagreement about contested interventions and the invocation or Rawlsian RE to resolve that disagreement, the burden of proof is on the proponent of PA to offer a such a process. Second, and of greater importance, is that fact that if their position rules out reasonable accounts of the moral life, then they must explain why that is the case and offer justifications for doing so.

These are important points, and they influence where the work needs to be done and who is to shoulder what responsibilities in the arguments over conscience in medicine. Some of these burdens are to be carried by the CD, at least by the proponent of PA, including the burden to offer a means of resolving or adjudicating disagreement. To close this section, and the chapter, consider a few final points on burdens of proof. Consider a common and simple case that highlights this challenge and the failure of the conscience dissenter position. Suppose two medical professionals hold divergent views about what counts as an appropriate reason for members of their profession. One might root their work in the Hippocratic Oath another in their experiences and formative training at an institution which did not have its students profess that oath. We might ask, is the best approach to resolving this tension to jettison one of these professionals or force all similar subscribers into a subfield? Both seem almost comical and are clearly not good faith responses to resolving disagreement, but this is, in actuality, the recommendation from proponents of PA. CDs, like Stahl and Emanuel, suggest that those not on board with the conscience dissenter position ought to be removed from the profession. In fact, Stahl and Emanuel claim that ought to either *become radiologists* or *leave the profession.*[59] Ignoring whatever it is that Stahl and Emanuel think of radiology as a subfield of medicine, how is it that this

[58] This critique is similar to Aristotle's point about levels of precision fitting the thing that one is considering (see Book I of Aristotle and Urmson [1980]), this might simply be too much philosophizing for the kind of view that is being offered.

[59] Stahl and Emanuel (2017, p. 1383). Further, it is unclear why radiology ought to be singled out as a place in medicine where one's conscience is not relevant.

conclusion follows from their discussion of the process of RE? Additionally, why should it be assumed that it is the medical professional (or in the argument about the place of conscience in medicine, the CP) who bears the burden of proving what is an appropriate reason to ground their practice? It is clear that CDs do not take seriously the sincerely and deeply held commitments of practitioners (a mark of discounting the integrity of physicians), which raises questions about the kinds of professionals they might seek out themselves and how attractive joining that group of professionals might be for prospective future physicians.

Thus, it is the proponent of PA who bears the burden of argument here. That said, the claim that is being argued for is not that conscientious dissenter positions, as a whole, lack merit. There do exist fitting concerns which motivate positions gravitating toward the pole of dissention. One of the more obvious of these concerns is the potential that respecting conscience in medicine would allow for protected professionals to violate standards of care, which are based on evidence and directed at the good of the patient. This is not generally the case, though there always exist outliers and those who would exploit a good ruling as a loophole for poor behavior or allow bad actors into play. It is important to (and for) the profession of medicine to uphold high standards of care, to care for patients in light of the best available techniques and the most recent advances in the medical sciences. Yet, disagreement among individual professionals is both possible and plausible, a state of affairs which is acknowledged by the proponents of PA but which they fail to properly explain. Additionally they fail to offer a fitting and defensible means by which to resolve or even adjudicate that disagreement, and—in the end—fail even to accurately describe the professional space that they have in mind for such adjudication. Jointly these failures are devastating because, as argued above, it is that very disagreement by which the CD (at least the proponent of PA) distinguishes between what professionals can reasonably and unreasonably refuse (that is, whatever is currently contested).

The aim of this chapter is to address professional disagreement. This topic is a key feature of the argument of the CD, articulated through the views of the proponent of PA, which was examined and found wanting. The burden of proof is on the CD to address a number of significant holes in their argument, including explanations of disagreement and a process by which to resolve it. Even more important is the need for the CD to offer and defend an account of the profession of medicine. I have done so by relying on the work of Buchanan and by understanding the normative implications of that account of professions. Part of the arguments of this chapter involved defenses, in a limited way, of the role that personal (including philosophical and religious) reasons might play in processes of adjudication. The arguments primarily focused on better interpreting Rawls' work, which the CD relied on, and showing what their positions would have to take on if they engaged that work in a more robust fashion. However, the role of religion in discussions of conscience is vast and deep. Religion is often attributed as the reason for disagreement and also the reason that disagreement cannot be resolved and that conscience should not be protected. I turn to this idea—the role of religion in the debate over conscience in medicine—in the next chapter, before turning to moral responsibility (Chap. 4) and, finally, to a discussion of medical acts (in Chap. 5).

References

Al-Shura, Anika Niambi. 2014. *Health Communications in Traditional Chinese Medicine Cardiology*. Amsterdam: Elsevier AP.

Aristotle, W. D. Ross, and J. O. Urmson. 1980. *The Nicomachean Ethics*. The World's Classics. Oxford (Oxfordshire); New York: Oxford University Press.

Brehany, John. 2011. "The Indispensability of Hippocrates." In *Hippocrates Is Not Dead: An Anthology of Hippocratic Readings*, edited by P. Guinan.

Buchanan, A. 2009. Is There a Medical Profession in the House? In *Justice & Healthcare: Selected Essays*, edited by A. Buchanan, 175–202. Oxford, UK: Oxford University Press.

Byrnes, Jeffrey. "The Problem of 'Core Moral Beliefs' as the Ground of Conscientious Objection." *HEC Forum: An Interdisciplinary Journal on Hospitals' Ethical and Legal Issues* 33, no. 3 (September 2021): 291–305. https://doi.org/10.1007/s10730-020-09425-5.

Cavanaugh, T. A. 2018. *Hippocrates' Oath and Asclepius' Snake: The Birth of the Medical Profession*. New York, NY: Oxford University Press.

Cruess, Richard L., Sylvia R. Cruess, J. Donald Boudreau, Linda Snell, and Yvonne Steinert. 2014. "Reframing Medical Education to Support Professional Identity Formation." *Academic Medicine: Journal of the Association of American Medical Colleges* 89, no. 11 (November): 1446–51. https://doi.org/10.1097/ACM.0000000000000427.

Edelstein, Ludwig. 1943. *The Hippocratic Oath: Text, Translation, and Interpretation*. Baltimore: The Johns Hopkins Press. Accessed at: http://guides.library.jhu.edu/c.php?g=202502&p=1335752.

Elliott, Carl. 2024. *The Occasional Human Sacrifice: Medical Experimentation and the Price of Saying No*, 1st ed. New York, NY: W.W. Norton.

Greenblum, Jake, and T. J. Kasperbauer. 2018. "Forget Evil: Autonomy, the Physician–Patient Relationship, and the Duty to Refer." *Journal of Bioethical Inquiry* 15, no. 3 (September): 313–17. https://doi.org/10.1007/s11673-018-9854-9.

Hafferty, F. W., and R. Franks. 1994. "The Hidden Curriculum, Ethics Teaching, and the Structure of Medical Education." *Academic Medicine: Journal of the Association of American Medical Colleges* 69, no. 11 (November): 861–71. https://doi.org/10.1097/00001888-199411000-00001.

Huemer, Michael. 2008. *Ethical Intuitionism*. Nachdr. Basingstoke: Palgrave Macmillan.

MacIntyre, Alasdair C. 2007. *After Virtue: A Study in Moral Theory*. 3rd ed. Notre Dame, IN: University of Notre Dame Press.

Mueller, Paul S., Sarah M. Jenkins, Katrina A. Bramstedt, and David L. Hayes. 2008. "Deactivating Implanted Cardiac Devices in Terminally Ill Patients: Practices and Attitudes." *Pacing and Clinical Electrophysiology: PACE* 31, no. 5 (May): 560–68. https://doi.org/10.1111/j.1540-8159.2008.01041.x.

Mueller, Paul S., Abigale L. Ottenberg, David L. Hayes, and Barbara A. Koenig. 2011. "'I Felt Like the Angel of Death': Role Conflicts and Moral Distress Among Allied Professionals Employed by the US Cardiovascular Implantable Electronic Device Industry." *Journal of Interventional Cardiac Electrophysiology: An International Journal of Arrhythmias and Pacing* 32, no. 3 (December): 253–61. https://doi.org/10.1007/s10840-011-9607-8.

Pilkington, Bryan C. 2020. "Treating or Killing? The Divergent Moral Implications of Cardiac Device Deactivation." *The Journal of Medicine and Philosophy: A Forum for Bioethics and Philosophy of Medicine* 45, no. 1 (January 10): 28–41. https://doi.org/10.1093/jmp/jhz031.

Pilkington, Bryan C. 2021. "Conscience Dissenters and Disagreement: Professions Are Only as Good as Their Practitioners." *HEC Forum* 33, no. 3 (September): 233–45. https://doi.org/10.1007/s10730-020-09395-8.

Pilkington, Bryan, and Charles Binkley. 2022. "Disproof of Concept: Resolving Ethical Dilemmas Using Algorithms." *The American Journal of Bioethics* 22, no. 7 (July 3): 81–83. https://doi.org/10.1080/15265161.2022.2087789.

Rawls, John. 1999. *A Theory of Justice*. Rev. ed. Cambridge, MA: Belknap Press of Harvard University Press.

Rawls, John. 2005. *Political Liberalism.* Exp. ed. Columbia Classics in Philosophy. New York: Columbia University Press.

Savulescu, Julian. 2006. "Conscientious Objection in Medicine." *BMJ* 332, no. 7536 (February 4): 294–97. https://doi.org/10.1136/bmj.332.7536.294.

Searcy, M. 2013 "White Coat, Greek Oath," The University of Chicago Magazine. Accessed at: http://mag.uchicago.edu/science-medicine/white-coat-greek-oath.

Starr, Paul. 2017. *The Social Transformation of American Medicine: The Rise of a Sovereign Profession & the Making of a Vast Industry: With a New Epilogue by the Author*. Updated ed. New York: Basic Books.

Stahl, Ronit Y., and Ezekiel J. Emanuel. 2017. "Physicians, Not Conscripts—Conscientious Objection in Health Care." *New England Journal of Medicine* 376, no. 14 (April 6): 1380–85. https://doi.org/10.1056/NEJMsb1612472.

Wicclair, Mark R. 2011. *Conscientious Objection in Health Care: An Ethical Analysis.* Cambridge: Cambridge University Press.

Wicclair, Mark R. 2021. "Commentary: Special Issue on Conscientious Objection." *HEC Forum* 33, no. 3 (September): 307–24. https://doi.org/10.1007/s10730-021-09458-4.

Chapter 3
Religion: The Red Herring, Leading Us Away from Moral Responsibility

The description of the debate over conscience in the preceding chapters underscored the importance of a workable account of disagreement for resolving this debate. A few different approaches to considerations of conscience emerged and led to the demarcation of two clusters of positions: one in which something other than a professional's conscience is normative for their practice and another which allows conscience to play a primary role. In carving up the argumentative landscape in this manner, positions in these clusters were described as gravitating toward poles. There are two advantages of relying on such a metaphor: first, it allows for more flexibility than lists of position suggests; and, second, it leaves open a space for positions in the debate which fall somewhere in the middle—where the magnetic force of either pole is insufficient to bring a position into its direct orbit.

One point made in the previous chapter, which is in need of further discussion, involves the kinds of influences on persons and the sources of their reasons. In some of the conscience dissenter positions surveyed in the previous chapters, much is made of the influence of religion on practitioners who appeal to conscience. In fact, it is most common in the debate over conscience in medicine to encounter a description of those appealing to conscience as failing to leave their religious convictions at the patient's door, as inappropriately foisting their religious views on patients, or as causing problems having to do with religion. In this chapter, and in arguments that appear elsewhere,[1] I argue that this is mistaken. This chapter houses a nonstandard and generally unpopular but accurate assessment of the role of religion in the debate over conscience in medicine. Before engaging that argument directly, given how previous versions of it have been received, it is important to set the stage and to clarify at the outset what is at issue, why, and where the argument will lead us.

The claim that is most often rejected—and rejected strongly—is that appeals to conscience in medicine have nothing to do with religion, religious belief, or religious practices. Two points, each for a different audience, are needed.

---

[1] Pilkington (2019).

© The Author(s), under exclusive license to Springer Nature Switzerland AG 2025
B. Pilkington, *The Medical Act: Conscientious Practice in a World of Dissention and Disagreement*, The International Library of Bioethics 114,
https://doi.org/10.1007/978-3-032-00048-4_3

First, for CDs, such as proponents of PA, if the argument proves to be sound, then focus on religion as the root of disagreement over conscience is wrongheaded. This opens up space for another candidate of focus, and I argue at the end of this chapter and in subsequent chapters that what should occupy that space and what is truly at issue is the role (and varying conceptions of) moral responsibility.

Second, those who understand themselves to be protectors of conscience but for whom religion plays a key role in that argument, will likely find an anticipated conclusion via very different steps. Such conscience protectors will likely disagree with the premises (or at least about the truth of some of the premises) of the argument in this chapter but find the conclusion acceptable. These clarifications are offered prior to the argument, as mentioned, due to some unanticipated receptions of earlier versions of the argument. One especially surprising interaction is detailed in the preface of this book. Recall from that anecdote that embedded within the heated tone of the critique and the post-lecture follow-up questions was an accusation that such views should not be expressed in the town in which the talk was given, as it was a "religious" town. The exchange illustrated that how philosophical arguments are received and how some topics, terms, or concepts—regardless of the conclusion of the argument which houses them—need to be offered in a careful manner. What was surprising about the critique that an argument, the result of which was that religion is a red herring in debates over conscience in medicine, should not be offered was that it attacked an argument for a shared conclusion. Even though the conclusion of the argument was that space must be carved out for those for whom religious faith or practice informed the content of their decisions and practical reasoning, it was not well received. This exchange, shared here at length, is intended to help situate the discussion of religious views related to conscience, highlighting the contentiousness of such discussions. Contrary to the CDs—and especially to proponents of PA, like Stahl and Emanuel—but also in contrast with some CPs, issues associated with conscience do not have much do with religion. It is true that on the account offered in this text, claims of conscience (whether they be to positively act or negatively to avoid involvement with certain practices or particular approaches to medicine) for religious reasons would—in many cases—be protected; however, this is because of considerations of moral responsibility and only derivatively of the religious content of the reasons of particular practitioners. Motivating reasons for particular practitioners might depend upon religious content but this is surely not the case for all professionals. This opening framing of the following argument is intended to assuage readers, who –like the religious leader in that town some years ago—might bristle at the suggestion that religion is a red herring, and be able to hear the argument for what it is, knowing that they concur with the conclusion. Similarly, proponents of PA or other conscience dissention advocates might see the argument as a new way to think about the debates over conscience and find in it shared premises and, ultimately, that the approach taken in this book (a stepping back from some of the deeply entrenched positions and most hotly contested points in the debate) will help achieve the aim of opening up new spaces for agreement.

3.1 Common Appeals to Religious Beliefs

Appeals to religious beliefs are common in discussions of conscience in medicine in a variety of places. They occur, as noted in the previous chapter, in some appeals to conscience relative to contested interventions (recall the discussion of RE and PA, and the critiques of it). They also arise in reflections on the professional identity or professional identity formation of physicians who might be motivated by such reasons or for whom such reasons influence how they move about the world. Appeals to religion also occur when a professional seeks to be granted an exception to some rule or norm outside the practice of medicine, but, given their location and the state of the world, it is deemed necessary for their practice. For example, vaccine mandates[2] have been on the rise since the COVID-19 pandemic, and such mandates are a common example of potential appeals to conscience that involve religion.[3] In fact, for some time there were three categories of reasons that one could offer in order to (hope to be) exempted from a vaccine requirement: health-related reasons, scientific progress-related reasons (something like "these drugs are too new" which was often in place only until the FDA approved the vaccine in question, at least in the US), and, finally, for religious reasons. After vaccines were approved, at least in the case of the COVID-19 vaccines, this left really only one option for those who wished to continue healthcare employment and to avoid vaccination: request an exemption on religious grounds. There is much to say about each of these cases, especially the latter, given the tribunals that were held, the epistemically poor interviewing procedures that took place at some institutions, and the generally not-well-thought-out reflections on conscience, religious practice, and medicine that transpired.

The rise in appeals to conscience and the apparent rooting of those reasons in religion appear to be on the rise, at the same time that we can notice an increasing trend in thought in healthcare, bioethics, and also in public discourse in line with CDs. The increase in conversations about conscience in medicine might very well be due, in part, to the COVID-19 pandemic, as noted above but it has gained new energy recently, at least in the United States, due in large part to a fraught socio-political context in which claims of conscience are often evaluated and determined to be one of two very different kinds of things: an assault on vulnerable persons or grave affronts to members of medical (as well as other health) professions.

In this section, and in different ways throughout this chapter, the role that religious claims play in the debates over conscience in medicine is examined. Many of the conversations, including those that are theoretical and philosophically—or sociologically—technical, as well as those that are logistical and practical, that touch on the role of conscience in medicine frame it as a debate over whether medical professionals should be allowed to conscientiously object to particular healthcare practices and, if so, on what grounds. This is a narrow framing of the relevant issues and it is one that becomes even more narrow on many conscience dissenter accounts. This is

---

[2] Some have argued that mandates led to increased administration of the vaccine; for example, see: Howard-Williams et al. (2022).

[3] For one example of this discussion, see: Jones-Nosacek (2024).

because, on those accounts, it is not only that some professionals are requesting to go against the grain but that they do so because of religious reasons and, many times, are attempting to foist those reasons on others. Wherever one falls in this debate, the narrowness and specificity of a CD's set of positions is clear. This is in fairly stark contrast with the way that issues related to conscience have been described in this text, which argued for respect for the conscientious practice of individual professionals, and gravitates toward a pole of conscience protection. In this chapter, the CD's narrow framing relative to the role of religion in the debate over conscience in medicine is scrutinized. It is to that framing that we now turn, in earnest.

This narrow framing is often articulated in two ways. One articulation describes appeals to conscience as inappropriate, unprofessional, unethical, or the like because it involves the foisting of a medical professional's personal religious values onto a patient. This foisting is not simply inappropriate, unprofessional, unethical, or the like, but a vicious violation of one of the referenced sets of norms (propriety, professionalism, ethics, et cetera) because it is a person in a vulnerable state—both in terms of the power and knowledge they bring to the conversation, as well as their (likely) sickened state—who is being treated in this manner. Because conscientiously objecting to some practice results in this kind of violation of patients, it ought to be banned (according to any of the referenced sets of norms), or so a common argument goes.

A second articulation describes appeals to conscience in a different light and, interestingly enough given the similarity of breadth, argues in favor of conscience protection. This second articulation goes something like this: conscientiously objecting to a particular practice which is viewed by the profession differently than the professional themself, is simply the exercising of their conscience and is an appropriate response to disagreement over matters of ultimate concern within the professional's profession, within the society in which the professional practices, or in light of the professional's deeply held religious beliefs. This second articulation shares features with the first and highlighting that similarity is important because it helps to locate the positive position taken in this text against CDs, as well as some CPs. Before turning to the more contentious claims about religion, further explication of the second articulation is needed. The conscience protector (for religious reasons) sees CO, like CO in wartime by pacificists, to be a perfectly acceptable way for a person to interact with others in light of their religious commitments.[4] The issues about which one might consider conscientiously objecting, on this view, could (and, in fact, world) be more numerous than the tight set of contested issues that Stahl and Emanuel reference, for example. Any workarounds, accommodations, or, more generally, responses would have to be in line with the religious practices of that objector.

It should be noted that though this narrow framing and its focus on religion is extremely prominent in the debates over conscience, the religious framing is not the only one that exists. Once the relevant features of the debate have been described,

[4] Here, the CP for religious reasons would need to take up arguments for the dissimilarity between CO in wartime and in medicine. In particular, they would have to respond to arguments like those offered by Stahl and Emanuel (2017).

it is argued that the root of disagreement over conscience in medicine is not religious claims but rather divergent (and sometimes opposing) conceptions of moral responsibility. The reasoning for this claim is that what is really at issue in debates over conscience is not religious claims, or appeals to religion, or that fact that health issues or interventions touch on issues or possess features which, due to the fact that health is a human concern and healthcare is a human practice, might also be impacted by religious practice. The issue at the root of the disagreement is the relationships between persons and the divergent (sometimes opposing) understandings of those relationships. If this contention is correct (and if the arguments of this chapter and—most especially—the subsequent one are correct), then the attention of ethicists working in medicine should be refocused from disagreements that appear intractable to working in places where a resolution now might be possible.

Before turning to the key point in this chapter, that religion is the red herring in the debate over conscience in medicine, points against and in favor of the inclusion of religion in medicine's discussions, practices, and the professional identity of its members will prove useful. Let us now to turn this topic by briefly surveying the landscape of common objections, as well as affirmations, of engaging religion in medical contexts.

3.2 Common Objections to (and Affirmations of) Religious Beliefs

In this section, a few objections and a few affirmations of religion in medicine are considered. This part of the discussion is intentionally framed in a broad and general manner so as to capture a variety of ways in which different practices, theories, rituals, and understandings of religion impact, influence, or are (potentially) bound up with medicine. I begin with the objections.

There are a number of reasons that one might seek to restrict the engagement of religion by the institutions of medicine or medical professionals. First and foremost, religious practice is often understood to be something very different from the medical sciences-supported, evidence-based practice of medicine.[5] Given the interest in emphasizing evidence-based practice and in the hard scientific nature of medical cures and interventions, there are any number of reasons that this distinction is supported. Religion and medicine, or so the critique goes, are two very different things and reasons that are persuasive (or even admissible) in one sphere are simply not relevant or acceptable in the other.

Secondly, not everyone goes in for religious practices and when they do, they are of a variety of kinds and connected to a variety of religious traditions; thus, it would be challenging to uniformly engage or even begin to describe what this

[5] This is not to reject claims of evidence in religion (or within religious traditions) or to even weigh in on the question of what constitutes evidence in such spaces. The point in this section is merely to highlight some reasons for and against interactions between religion and medicine.

should look like. Consider again the aforementioned, albeit narrow, reduction of considerations of conscience in medicine to CO. These challenges may help to make greater sense of that framing. Addressing the diversity of religious practices is not easy if one's aim is to offer a consistent policy for the administration of healthcare and the relevant logistics in delivering that care. However, if one sets policies that simply reject engagement with religion and then one carves out potential exceptions on a case-by-case basis with strict requirements about promulgation, timing, and other obligations, potential service interruptions might be minimized.[6]

Thirdly, it may be philosophically or ethically challenging to justify such practices, at least under some commonly appealed to philosophical approaches. Suppose, to continue with the contractualist line of thinking described in the last chapter, that religion openly played a role in medicine, how would the diversity of religious practices and emphases be engaged? Consider, for example, the idea that professionals ought to offer reasons to others that they would consider for themselves and then determine acceptability of a practice; that is, suppose it is the case that (given the narrow CO-focused framing) physicians need to justify to others (e.g., health systems, colleagues, or patients) why it is that they are conscientiously objecting to something, whatever that something is. Justification is a fair and important feature of moral relationships and given the framing by the CD of conscientious practice as CO, it would be the physician explaining themselves to others. One commonly appealed to contractualist position, offered by T. M. Scanlon,[7] explains this justification in terms of reasonable rejection. That position illustrates well the challenges that a CD may wish to articulate if justification to others is prioritized. In *What We Owe to Each Other*, Scanlon argues in favor of a particular kind of contractualism.[8] Scanlon, whose account prizes whether others could reasonably reject a principle as the standard for moral evaluation, writes, "An act is wrong if its performance under the circumstances would be disallowed by any set of principles for the general regulation of behavior that no one could reasonably reject as a basis for informed, unforced general agreement."[9] This emphasis on reasonable rejection shows how large the role that justifiability to others plays in Scanlon's consideration of practical reasoning and it is a fitting lens through which to consider requests of CO and their potential ethical and philosophical complexity. On Scanlon's view, we must recognize our friends as persons with "moral standing." Something similar might be said to physicians wishing to avail themselves of conscience protection, which is—at least on the CD's view—to request respect for their CO. Given Scanlon's emphasis on justification, it is no surprise that those to whom justification is owed

[6] It should be noted that this is only the case if we are considering a CD position that allows for space for some kind of conscientious practice, such as Stahl and Emanuel's allowance for currently contested interventions. Not all conscience dissenter views afford such allowances.

[7] Scanlon (2000).

[8] The details of his overall account and situating it within the broader contractualist literature are both beyond the scope of this book and not necessary for the current argument.

[9] Scanlon (2000).

is of the utmost importance. Consider, again, three possible answers to the question of *to whom* conscientious objectors might owe justification: the health system, colleagues, or patients. Scanlon, as his work is a more broadly applicable instance of moral theory, focuses on the class of creatures that can be wronged by our actions. In that vein, we might also add to list: members of the community in which the health system operates, those who are close to or impacted by changes in schedules of and demands on other clinicians, those who are close to or impacted by intervention options for patients (including even society, at large). Let us consider these three, though the arguments could be run more broadly. Scanlon, himself, claims that those to whom we can stand in relation are those that should be considered in thinking about justification or, as he puts it, those who fall within the scope of morality.[10]

The benefit of Scanlon's form of contractualism, at least according to his own work, is that it answers the two problems presented to any moral theory: the priority of right and wrong over other values and the special problem of importance (that is, that not responding to moral reasons is worse than not responding to other kinds of reasons). In analyzing the role of CO through this lens, one can see how Scanlon's argument and focus on justification and reasonable rejection helps to capture and explain further the conscience dissenter position, as well as the complexity of the relevant issues. I take each point in turn.

First, let us focus on patients, colleagues, and the health system in which care is embedded. Given the vulnerability with which patients present, one might read the priority of moral questions in Scanlon's work—similar to the priority of profession-based questions in the conscience debate. If it is of little or no cost to accommodate the conscience of a medical professional, relative to patients, colleagues, and systems, some CDs might find this acceptable. However, the norms of the profession, guided by altruism, and directed toward patient-centered care, place the considerations of patient vulnerability in the moral category and, as such, this always takes priority over other considerations. It takes priority over, to continue to use the language of prominent theories of CO, the sincerely and deeply held beliefs of persons offering or contributing to that care. Relatedly, with respect to colleagues and health systems, not responding to requests for accommodation, in light of the CO framing, is of much less importance than not responding to patient needs and demands, almost as if there were no such requests for the protection of conscience.

The rhetorical force of Scanlon's position is powerful, especially if one considers potential conversations (a common first step in attempting to adjudicate disagreements) between professionals, systems, and patients. One can imagine how such conversations would go in the instance where a patient requests a service, a colleague does not wish to shift cases or take on additional burdens, or even health system administrators do not have time to consider the reallocation of duties: could they reasonably reject those requests? CDs would firmly and emphatically say yes.

---

[10] Scanlon settles on those beings that are capable of making judgments. See Scanlon (2000). It should be noted that in this group he includes those who have died and those who have not yet been born. It is still wrong to harm those in other categories; it is just not the same kind of thing as harming someone within the scope of morality.

Returning to Scanlon's moral theory, he argues that ignoring moral reasons puts a gulf between the person who does so and others in a way that not responding to other reasons does not. This is illustrated by the shrill tone of the debates over conscience. For Scanlon, this gulf illustrates morality's importance; for considerations of conscience, this illustrates its lack of importance in the eyes of some, the significant priority that is given to what falls within the bounds of the profession for others, or the distaste of appeals to religion in these contexts for still others (and each of these is not mutually exclusive). Morality, for Scanlon, is prior to other values because we need to justify our actions to others in order to have good relationships with them. Similarly, the CD might argue, professional norms are prior to personal norms, integrity, and one's own self-conceptions and one's own values.

However, what this second foray into the contractualist landscape illustrates is not a decisive argument in favor of the conscience dissenter position, nor a "win" for that "side" of the debate over those who prize conscientious practice or those who find religion and respect for personal religious views to be important, even in health contexts and even if they are the views of those offering care. What reflection on Scanlon's notion of reasonable rejection actually does is that it illustrates the complexity of the situation. This complexity includes the depth of disagreement and the importance, for those thinking about the role of conscience in medicine, of addressing and resolving that disagreement. This is because, at least in part, the multi-directionality of Scanlon's powerful question. One might just as well ask whether requests for the maintenance of one's moral integrity could be reasonably rejected; that is, one could run the conversational thought experiment in a different direction or with different terms and one might arrive at different answers. Suppose those administering a health system fail to allow space for conscientious practice in their policies. Suppose further that a physician, who is part of that system, requests this space or conscientiously objects to certain interventions. When pressed, what if the physician asks the administrators whether they can reasonably reject the physician's request?

Scanlon's actual formulation engages general principles and a wider population, who are uncoerced and informed, to determine what could be reasonably rejected. Thus, let us consider the above to be a Scanlon-inspired thought experiment, which aimed to do this in the practical setting described. In that context would it be possible to be sufficiently informed and uncoerced to make such a determination? The usual answer in medical ethics is "yes," as long as certain conditions obtain[11]; however, cases that involve religion (or any set of complex beliefs that involve meaning, histories, traditions, and rituals)[12] require a more nuanced analysis. This is because, at least in part, of the challenge of sharing sufficiently full information in order to offer an explanation that is persuasive enough to serve as justification across

---

[11] Consider the Belmont Report's restrictions on undue influence, unjustifiable pressure, and coercion, for example. (Department of Health, Education, and Welfare and National Commission for the Protection of Human Subjects of Biomedical and Behavioral Research. The Belmont Report (2014).

[12] See, for example, the arguments in Chap. 4 about autonomy-affirming practices and the challenge of taking up another perspective. This is not a novel point, but it is an important one.

conceptual languages, traditions (in the MacIntyrean sense[13]), and practices. These sorts of translational issues are notoriously challenging, leading to loss of content, transmutations, and various forms of confusion.[14] Without delving too deeply into these issues, we might ask: how much information (and how much time would it take to acquire the necessary information) is needed to sufficiently understand a religious system so as to judge a claim rooted in it? Additionally, what kind of experiential knowledge would be needed? This may be another point in favor of the CD who demands that only public reasons be offered, if nothing else they (at least purport to be) understandable by all. The cost, of course, is that even that category of reasons does not exist in a neutral space, and so this attempt at resolution appears to only move the issue to a different place in the argument, not resolve the disagreement.

With this background it is high time to consider not only the philosophical resources and potential reasons the CDs reject religion, even if the issues are complex, and turn to what one especially vocal set of CDs and critics of religion and medicine actually argue, Stahl and Emanuel. Addressing their claims here will be useful, as their position was engaged in the last chapter and the reader will be familiar with some of the relevant conceptual resources, argumentative strengths and shortcomings, and aims of those arguments. It is to those arguments that we now turn.

Recall from the discussions in the preceding chapter that Stahl and Emanuel argue that medical professionals are bound by the norms of their professional, not their conscience, and that this is due to the fact that the primary interest of the medical profession, and so of medical professionals, is "to promote the well-being of patients."[15] The reader will also recall that when disagreement (at least with respect to conscientious practice or the more narrow, conscience-dissenter-preferred CO) arises within the profession, it is understood to be caused by the invocation of personal religious convictions and that such disagreement can be adjudicated through the process of RE. For their arguments to be persuasive, both the appeal to religious claims and the adjudication process must stand up to scrutiny. However, as we saw in the last chapter, the latter does not. Stahl and Emanuel's employment of Rawlsian RE fails to be sufficiently Rawlsian and sufficiently defensible, given their aims and the basics of Rawlsian political theory. However, to the point of this chapter, Stahl and Emanuel's position captures some very commonly articulated views on the subject of conscience in medicine, as it relates to religion, and so a more careful examination is warranted. It may be the case that their arguments about religion are more persuasive. They note, in referring to RE, "This dynamic process establishes professional obligations for health care providers regardless of their personal beliefs"[16] which they clarify, noting, "In a professional context, personal religious convictions

[13] For more on traditions, see Chap. 4 of this text; for a full articulation, see MacIntyre (2007) and MacIntyre (2006).

[14] For another discussion of this issue see: Pilkington (2018). For an historical analysis of the related issues in bioethics, see: Evans (2014).

[15] Stahl and Emanuel (2017).

[16] Stahl and Emanuel (2017).

are secondary."[17] They demonstrate this with the following set of situations, which are all variations on commonly discussed medical ethics cases, "Thus the Jehovah's Witness surgeon cannot refuse to allow blood transfusions during surgery. The Jewish pharmacist cannot withhold pills that are made with nonkosher gelatin. The Mormon nurse cannot refuse to treat alcoholics."[18] As noted in the previous chapter, there exist professional norms in medicine and as a result of those norms, medical professionals must leave religious belief at the patient's door. It is interesting to note that proponents of other positions in the debate, even some of those who seek to protect the consciences of medical professionals, accept the claim that religion is the root of disagreement. These CPs diverge from the dissenters because the former believe that religious belief is so important to persons that carving out space for conscientious practice is justified. As noted above, even on a Rawlsian account, space for religious practice (given its importance to persons) is admissible as a public reason; one is barred from offering religious reasons not from offering public reasons that take seriously the importance of religious practice to persons and the deep influence that religious belief can, and does, have on the comprehensive doctrines of citizens.

This claim is mistaken. It is not a mistake to note the importance of religious belief and practice in human life, nor to notice the complications that arise in attempting to respect diverse sets of values (with origins that can make explanation challenging and complicated), nor even to be tempted by the apparent consistency of a set of professional norms that might make all medical professionals act in exactly the same manner. The mistake is to think that debates over conscience exist because of religion's introduction into medicine. This is because religion is a red herring in this debate. Disagreement arises, contrary to most interlocutors on the subject, not because a physician possesses religious beliefs or that they make religiously rooted claims of conscience, nor even if their values are properly described as religious. Rather, disagreement arises because of beliefs, claims, and values associated with how persons interact with others. About this, Scanlon's position is both right and illustrative: ethics is about relationships.

3.3 Why Religion is the Red Herring (and Some Objections and Responses)

In this section, it is argued that religion is a red herring in the debate over conscientious practice. Instead of religion as the root of all divisiveness, I argue that the disagreement is actually due to differing accounts of moral responsibility. To begin this argument, I take up the examples employed by some CDs—and quoted above—and analyze them. After unpacking some of the built-in assumptions and playing out what the cases would look like in practice, it becomes clear that religion is not the issue that it is made out to be. In the aforementioned examples from Stahl

---

[17] Stahl and Emanuel (2017).

[18] Stahl and Emanuel (2017).

and Emanuel, three members of different health professions, each with a different religious affiliation is—to play out the example—denied an opportunity to practice conscientiously (or on the conscience dissenter view, to conscientiously object). In fact, the cases are merely a line, and stated in a way that suggests a certain felt rhetorical force is anticipated. It is almost as if the writers think that it is obvious to all that were such space (or accommodations) requested, it would be wrong to do so, and the requests would be obviously denied. The CDs are correct in this assumption; the requests would be balked at in most contexts and, largely, with good reason. This is because the cases themselves are constructed in such a way as to obscure—perhaps unintentionally—the deeper issues, issues which are more challenging to address but also which do not allow for persons of religious faith to be scapegoated in a complex and murky world at the nexus of personal identity, professional identity, and the care of others. Let us turn to those examples directly.

The three cases introduce the reader to members of the Jehovah's witness, Jewish, and Church of Jesus Christ of Latter Day Saints (LDS)[19] communities. A surgeon who is a practicing Jehovah's witness and does not want to use blood products in an operation, a pharmacist who is Jewish and does not wish to stock pills made with nonkosher gelatin, and a nurse who is a member of the LDS church and does not wish to treat alcoholics. One of the features of these cases that makes them interesting, and adds some of that aforementioned rhetorical force, is that these are all *almost* classic cases in medical ethics. Whether to transfuse the patient in need of blood who refuses it but lacks capacity is so common a case in medical school ethics sessions it would be surprising if a physician practicing today had not had a lecture on it. Those cases, almost, if not always, feature a character who is a Jehovah's witness. The twist in the case under consideration is that it is not the patient (or the person speaking for the patient who is a Jehovah's witness but the medical professional, the person offering care). These two features—a classic medical ethics case or issue coupled with a character twist—could bring with them the force of a reader's assessment of the other (the actual common case) to the current case, but—it is important to note—these are not the same cases.

Further, adding the additional cases to the mix, the pharmacist and the nurse, there is also something awry. What is common to all three cases is that each character, according to the CD's assessment, should engage in activities with respect to other persons which they presumably would not choose for themselves, and the reason that they would not make such choices is due to their own personal, religious-based reasons. One can fill in features of the argument from the previous chapter at this juncture. Personal religious reasons ought not to be foisted on others, especially when those others are in vulnerable states and even more so when the person doing the foisting is a member of a profession that is partially constituted around claims of altruism. (Part of the reason that these cases are so persuasive, part of the reason they carry such rhetorical force is because of this foisting: the imposition of reasons that are not one's own is not something that anyone would take lightly, let alone patients

[19] LDS is employed as an acronym for the church, and its members, which Stahl and Emanuel refer to as "Mormon."

in vulnerable states.) Such reasons are not admissible on Rawlsian grounds, notes the CD. Let us take up these two claims, in reverse order, but, first, put to the side the previous critiques illustrating that PA does not do justice to Rawlsian political theory. It is true, as explained above, that the reasons offered by the characters in these cases would not be admissible on a Rawlsian account, as they would be part of a comprehensive doctrine. That said, what they would not be admissible to is a conversation about constitutional essentials, in light the requirements Rawls places on public reason.[20] The concern the reader should have here, which brings us to the first claim, is not that some reasons would be admissible and others not; it is that such a conversation only arises if the character in these cases offer such a reason and the examples are not representative of appeals to conscience in medicine! The characters in these three cases may very well be motivated by the content of those religious reasons relative to blood products, gelatin, and alcohol for themselves but it is not clear that practicing Jehovah's witnesses, Jews, members of the LDS church, or persons of religious faith, generally, view their reasons as motivating (or as reasons that ought to be motivating) for others. Therein lies the twist in the classic case. The problematic nature of these three appeals to conscience may not, in most cases, even get off the ground.

To demonstrate this, consider the cases. Merely because a nurse, who is a member of the LDS church, did not drink alcohol themself does not mean that they would refuse to treat a patient who did imbibe or even a patient who suffered from alcoholism or another substance use disorder. Were a surgeon also a practicing Jehovah's witness, they might refuse blood products, even if they were necessary for their survival, but it does not necessarily follow from this that they would refuse to operate on a patient whose own survival required the use of blood products. The Jewish pharmacist might refuse to use pills made with nonkosher gelatin themself, but they might not object to the stocking of such pills in their store, if properly labeled and with alternatives available for those with similar faith commitments to their own. These cases fail as illustrations of appeals to conscience, but in so failing they help to highlight the key point of this book: disagreement over conscience in medicine is not about religious claims; instead, disagreement arises when professionals understand particular acts of which they would be a part as wrong. It is only if the Jewish pharmacist declines to stock the pills, and such pills are not easily obtainable at the pharmacy down the block that the debate over conscience in medicine—framed by the CD in terms of CO due to personally held religious claims—gets noticed. What is actually at issue in these cases, and in the debate over conscience in medicine as a whole, is the right account of the relationship between persons and how that account impacts the actions performed or omitted in healthcare spaces; that is, it is a debate over moral responsibility. What the conscientious dissenter does, as noted in other chapters and elsewhere,[21] is to beg the question against the CP by assuming answers to an open question about reasonable accounts of the moral life. The CD (both proponents of PA,

[20] An entire book could be, as many books have been, written about public reason. That is not the focus of this text. For a discussion of constitutional essentials, see Rawls (2005).

[21] See Pilkington (2019). For another discussion of these issues.

like Stahl and Emanuel, and others) assumes that accounts of moral responsibility, in particular, accounts in which an agent might be held morally responsible for the results of their actions (even if the consequences of those actions affected another person or persons to a much greater degree) are not reasonable.

If this is correct, then it is not religion or religious belief or practice at the root of disagreement over appeals to conscience in medicine, but accounts of moral responsibility. Focusing on religion, as (at least some) CDs do (recall Stahl and Emanuel's claim: "In a professional context, personal religious convictions are secondary."[22]) is a red herring. Yet, members of both sides of this debate make this mistake: some CPs agree with the conscience dissenters here, they simply seek protection for their (or others' religious beliefs and practices). However, further examination of the cases, despite their religious framing, illustrates that the issues generated are not really about religion. Consider, again, those cases. The CD would not attempt to convince the surgeon to allow blood products to be used in a surgery that they might undergo themself, but rather not to deny those products to others: thus, the issue is not, at its core, with particular religious convictions, but with accounts of action. The concern of the three health professionals in question is with their being bound up (in certain kinds of ways) with acts they understand to be wrong. That is the basis of the disagreement. Again, it is not that some claims have religious content or even that some medical professionals are members of religious traditions, or even that some physicians' self-conceptions are deeply informed by being a part of those religious traditions. It is only when one of those professionals holds an account of moral responsibility that highlights the interconnected nature of persons that the issue arises, and disagreement is generated.

To demonstrate this more clearly, let us return to the cases. Consider the nurse. As a practicing LDS member, the nurse would not drink, but why think she must consume alcohol in order to treat a person with a substance use disorder? To the contrary, the reverse might be true: it may be that that concerns about the consumption of alcohol, the disease of alcoholism or substance use disorders more generally would be a sensible place for care by someone with this orientation. For a stronger case, consider rehabilitation for substance use disorders that is more extensive. Consider a psychiatrist who is a practicing member of the LDS church and who is approached to treat a potential client suffering from substance use disorder. The CD's analysis only works if, in this iteration of the case, the standards of psychotherapy required that therapists and patients engage directly in the actions which were the source of the issue together to shed light on the basis of the relevant psychosis. In this version of the case, the LDS psychiatrist would object; however, it is worth noticing that the issue is not with the psychiatrist's LDS-based beliefs about alcohol or, to consider the other cases, the Jew's views of nonkosher gelatin, or the Jehovah's witness' beliefs about blood. The CD's almost exclusive focus on religious reasons leads examination of these issues in the wrong direction; the real issue is that they reject—and without argument—an account of moral responsibility.

[22] Stahl and Emanuel (2017).

If the arguments in this chapter are sound, our gaze may be refocused on how to adjudicate disagreement, but not disagreement over religion in medicine, in general, and over one person imposing their deeply held religious beliefs on another, in particular, but rather disagreement about conceptions of moral responsibility. If this is right, then restricting the conscientious practice of medical professionals who engage (a) religion is not a fitting resolution to this disagreement because the root of the disagreement is not religiously based. Rather, an examination of moral responsibility is needed. Before turning to that topic in the final section of this chapter, a few objections to this proposal ought to be considered.

3.4 4: Religion, not Moral Responsibility: Objections to a Reframing

The arguments of the preceding section illustrated that the deep disagreement that exists in debates over conscience are about moral responsibility and not religion. There are a few important objections that call out for responses, especially due to the fact that the position is not commonly held and requires a reframing the debate's focus. Consider three of these objections.

First, reframing the debate over conscience in medicine surely has its benefits. It opens up space for potentially fruitful dialogue, as the new options for agreement might arise around different conceptions of moral responsibility, as opposed to the seemingly intractable exchanges and entrenched proponents of conscience dissention and conscience protection that exists in the current debate. It also accurately reflects where the real issues exist. However, this does not mean that there are no costs to such a reframing. One (potentially high) cost is that shifting the focus away from religious practice, or the broader category of religion, risks jettisoning some protections for conscience which were—even if in a limited manner—in place. One might argue that religious beliefs are foundational to the lives, practices, and even the ethics of many persons, including some physicians.[23] Thus, one might worry that moving away from protections afforded persons—even if missing the mark in the debate—raises concerns about the limiting of religious freedom. This critique encompasses two lines of criticism. In the first place, it relies on a basic disagreement with the contention: that conscience debates are not truly about religion but rather about something else which religion touches. This belief, or something akin to it, is what appeared to be motivating the religious leader who rejected this analysis so vociferously some years ago and detailed in the preface. In the second place, it is a worry that the arguments of this book might not be successful in their positive veins but only in their negative ones: that is, one might take the arguments against the inclusion of religion as a way of bolstering a conscience dissenter account without properly attending to the

[23] For considerations of religious bioethics see, for example: Engelhardt (2000), as well as Engelhardt (2017), and Smith Iltis (2010).

recommendations associated with moral responsibility, thus—potentially—gutting protections already in place and undermining the project of safeguarding conscience.

This is a serious objection, and merits further thought. Because it includes two distinct considerations, it is owed two responses. First, the claim that religious beliefs serve as a foundation, informing (and forming) the rich moral lives of many persons, including medical professionals, neither conflicts nor is in tension with the judgement that key issues in the debate over conscience in medicine are not religious. It is not inconsistent to hold the view that religious beliefs, practices, and traditions are deeply meaningful to devout and faithful persons, while also holding the view that what is at issue in the disagreements at the heart of this book have to do with how persons perceive themselves to be interacting with others. To be clear, the claim at the heart of these arguments is not that religious beliefs, practices, and traditions do not matter, nor that they do not matter to medical professionals, nor even that there is not a place for them in care for the health of persons. Devout medical professionals are surely formed, in part, by these practices, beliefs, and traditions. In fact, it may very well be the case that a devout Muslim, Jew, Hindu, or Christian finds the ethical moorings of their professional practice in their self-conception as a religious person. It might even be that such a professional was originally attracted to the work of the medical profession due, in part, to that faith. A deep love of neighbor coupled with an understanding of the importance of health could easily lead someone to medicine as a vocation. It does not follow from this account of religion in medicine that disagreements about opening up space for conscientious practice in that profession are not, at their core, actually about conceptions of moral responsibility. It is not even in conflict to hold that one's preferred conception of moral responsibility be informed by features of the religion that one subscribes to, so the connection may be a close one. However, this need not be the case, as was demonstrated above through an examination of the three cases: the Jewish pharmacist, the LDS nurse, and the Jehovah's Witness surgeon. To further support this contention, consider a case in contrast. Imagine a physician who holds no religious views or finds them to be untrue or unknowable—an atheist or an agnostic. Suppose that this physician is reflecting on the legalization of a new MAID law (one of the contested interventions, mentioned by Stahl and Emanuel) and suppose further that the law were to require physicians to aid in the death of patients, with no opt out possibilities or conscientious practice carve outs. This physician might very well argue against the law, supporting lobbying efforts to change it. Suppose they do so by means of arguments that root their analysis in a conception of the medical profession as a healing profession: doctors ought not to be involved in aiding in the death of patients, for to do so is to fail as healers. Consider a second physician, who is a devout Christian and, as such believes that to kill is a violation of the fifth commandment, which forbids unjust killing, and to fail in the eyes of God. What if this Christian physician, holding firm to this belief and knowing that they would never knowingly transgress God's law, would never themself seek their own death. However, suppose, this Christian physician believes that they are distinct from others and that reasonable persons can disagree about conceptions of the good—having read some Rawls in college—and, as a result does not have an issue prescribing for MAID. This second physician may still think it abhorrent to end one's life but does not see themselves as being bound up in the right sort of way with the state of affairs that obtains in MAID situations, and so they are not connected, that is to say, not morally responsible for any wrongdoing.

If these caricatures reflect truth, then they help to illustrate the argument that has been building: it is not religion, but moral responsibility that matters in debates over conscience. I do not wish to argue over the details of the religious doctrines practiced by the second physician, nor to investigate the atheist's or agnostic's interpretations of healing. The key point illustrated by these cases is that if they are plausible examples of medical professionals, they illustrate—again—that the root of disagreement is not that some medical professionals are religious and others are not, it is that each medical professional understands differently what their interactions with others mean, that is, they possess different conceptions of moral responsibility.

A reply to the second point to be drawn from the first objection is also called for, though this reply might be unsatisfactory for those who harbor concerns about the potential loss of current protections of conscience. There are good reasons to be concerned with limitations on religious freedom; however, it does not follow from these concerns that my assessment of the debate is inaccurate. A proper response to these concerns is to remind those who might take some arguments from the book and not others to, please, consider the arguments in their entirety. It is not that the arguments fail to do justice to religious liberty or similar concerns, it is simply that when it comes to debates over conscience in medicine they are not what is really at issue.

A second objection, one that is more concerning than the first, is that many conceptions of moral responsibility have a basis in religious beliefs, practices, and traditions. In light of this, so the objection goes, the problem has only been moved to a different place; it has not been resolved.

This is a serious objection; if the problem has only shifted locations in the argument, then the argument has failed. The objector rightly notes that some accounts of moral responsibility might very well be religious in nature or are, at least, informed by religious traditions, potentially in such a way that the two cannot be separated. If this is correct, if conceptions of moral responsibility are so closely bound up with religious traditions, then religion is not a red herring—as has been argued—but is, in fact, the central issue to be discussed in debates over conscientious practice.

Thankfully for the status of the argument, the objector is mistaken. They are correct that religious beliefs and traditions inform some accounts of moral responsibility, and it may even be (and, in fact, surely is) the case that how some medical professionals understand their interactions with others is due to their religious beliefs. Yet, this does not mean that the two are inseparable. Religious content may inform some physician's conceptions, experiences in emergency departments another's, and a close reading of the Hippocratic Oath or the Physician's Pledge, a third. The objector's critique does not bear on the key issue, but rather—and this is a topic addressed in subsequent chapters—it highlights one set of conceptions of moral responsibility. To illustrate the aforementioned separability, consider the example, offered above, of the agnostic or atheist physician whose conception of moral responsibility yields that they are responsible for the death of the patient in a case of MAID, but recall that nothing religious contributes to that content. To the contrary, it is a secular accounting of medicine as a healing profession which does the work. And this line of argument works for a variety of caricatures. Suppose two physicians, one of religious faith and the other not, find natural law accounts to be persuasive. Suppose they both spent time pouring over the philosophical work of Thomas Aquinas, and they both have adopted Aquinas' interpretation of action and of moral responsibility. Let us suppose further that the first physician does this for religious reasons: they are a practicing Catholic, they find Aquinas to be a clear and thoughtful theologian, and his saintliness was what first brought them in contact with his work. Let us suppose that the second physician has adopted Aquinas' philosophical worldview because, having studied philosophy in college, they found Aquinas' account of action and how it conceptualized the interactions between persons to resonate with their experiences working within communities. This second physician has no particular religious views; they are agnostic about such things. Can we make sense of these cases? Do these physicians appear reasonable? The answer to both questions is "yes." This is because religious content can inform a person's conception of moral responsibility, but it need not. Additionally, regardless of a medical professional's particular religious affiliation or practices, they might concur with others of differing faiths or of no faith at all about the right conception of moral responsibility.

A third objection takes up considerations of religious liberty connected with the first objection, finding my responses to the latter inadequate. Unimpressed by my response to the first objection, the third objector roots a call for retaining the resources in the social and legal history of the United States (and, potentially elsewhere), arguing that that history of the protection of religious liberty allows for the specification and adjudication of religious liberty claims within legal systems. To move away from these protections with legal power behind them, even if deemed by some CPs to be insufficient, is foolhardy. First, it leaves fewer resources to protect conscience. Second, its replacement—moral responsibility—is a poor substitute because, even if my arguments are right, accounts of moral responsibility themselves cannot be adjudicated easily.

These are strong objections rooted in practical concerns. Consider each in turn. First, the appeal of relying on safeguards for religious liberty that have already been carved out in the law is understandable. It is true that, given the social and legal history (at least in the U.S.), the court system may be set up to adjudicate such issues. Yet, both of these claims can be granted and it can still be true that the debate over conscientious practice is a distinct matter. The objection is, in actuality, a point about strategy. It is an important one, but it does not bear on the conceptual analysis. Second, one might argue that, though the analysis is persuasive, it only trades one sent of intractable disagreements for another—fights about religion are replaced by fights over conceptions of moral responsibility. This second objection is stronger than the first, and it is a powerful objection for at least two reasons. It is powerful, in the first place, because of the influence of liberal political theorists, such as Rawls, and the clear influence that their (and especially his) work has had on CDs. Rawls considered not only controversial religious doctrines, but also philosophical positions, in what he understood as a person's comprehensive doctrine. A second reason that the argument is powerful is that one might find accounts of moral responsibility to be complex, nuanced, and requiring a good bit of philosophical training to parse out. One might claim that if full theories of action or of communities of persons within whom interactions might exist must be engaged, then this proposal is (almost) a nonstarter, as it is just too much to take up in order to resolve the debate.

Though these objections are both strong and on target—as opposed to the aforementioned objection—neither is sufficiently persuasive to derail the project at this juncture. Replying adequately to the first requires extensive engagement with the Rawlsian corpus. This is done in the preceding chapter, with discussions of key Rawlsian concepts such as RE, comprehensive doctrines, and public reason, as well as conceptions of equality and freedom. If more is needed to respond to the objector, it is beyond the scope of this text. That said, such an accounting should come from CDs both because they are the ones appealing to Rawls, but also because the intractability of the disagreement need not only be resolved through Rawlsian political philosophy. Second, conceptions of moral responsibility, as well as theories of action might be complex and nuanced, but there is no reason to think that they are more so than religious traditions, practices, or beliefs. Thus, the shift that has occurred has not merely replaced one intractable disagreement with another but opened up a new space for the actual debate to take place. To that end, the arguments of this chapter, if they are sound, have actually advanced the debate and done so in at least two ways. This is because, first, they have shown where the real disagreement is—in differing, if implicit, accounts of moral responsibility—and shifted focus from a valuable but peripheral component of the debate: religion. Second, the arguments of this chapter also point a way forward, beyond the current morass. If the debate is actually about moral responsibility, then what CDs must argue for is the claim that those who oppose the stance of a professional association (of which they are a member) related to a conception of moral responsibility, must hold an unreasonable account of moral responsibility. That is, the CD must claim that those who understand themselves to be in relationship to other members of their community in such a way that they share in the responsibility for how that community member's life goes would hold an unreasonable account of moral responsibility and, in turn, of the moral life. This is not a tenable position to hold, but the CD is owed a positive characterization of moral responsibility and an argument for it, as well as a discussion of the objects of those conceptions of moral responsibility in medicine: medical acts. I turn to the former in the next section and to the latter in the subsequent chapter.

Before turning to those subjects, a short summary of the argument thus far may prove useful. In the first chapter, the overall landscape of the debate over conscience in medicine was addressed, terms were clarified, groundwork laid, and two poles toward which positions in the debate gravitate were identified: conscience dissention and conscience protection. The second chapter houses a deep dive into disagreement and, after careful analysis, found the conscience dissenter position wanting, as it could not properly account for that disagreement. In this third chapter the mistaken emphasis on religion has been the target, with the central contention being that the focus on religion in discussions of conscience is both common and a common mistake. The conflict between CDs and CPs comes apart in their accounts of moral responsibility, not in terms of the content of particular religious traditions or practices, even if those bear on such accounts. It was also emphasized, given potential and aforementioned critiques, that it does not follow from these arguments that religious liberties should not be safeguarded nor do these arguments deny that religious traditions are genuinely meaningful to persons. Rather, the root of disagreement in the debate over conscience in medicine is not religion but accounts of moral responsibility. Attending to this conflict will help orient our discussions of the role of conscience in medicine in a more productive direction. Now, we can turn to moral responsibility.

3.5 Moral Responsibility: A Positive View

In this final section of this chapter I offer a brief sketch of what I mean by moral responsibility and what I mean by invoking it in the context of debates over conscience in medicine. Much of the argumentative work to show that moral responsibility, and not religious belief, is at the heart of the disagreements in this space has been offered in the preceding sections and chapters. This has been done primarily through reflection on examples and thought experiments, which have demonstrated that the crucial issue is how persons understand themselves to be bound up with one another and for what they might be responsible, in light of that engagement. This is important because if it is reasonable for a person to view themselves as, in part, responsible for the outcomes of their actions in certain contexts, then it is the case that (at least) some deference ought to be afforded them, in light of that responsibility. If I can be held responsible for your remaining sick, or dying, or getting well, or flourishing, then I ought to have some say in whether I am obligated to do so or, at least, it ought to be recognized that we are so connected.

At this juncture of the argument, it will be important to clarify what is meant by moral responsibility, to distinguish it from other concepts, and to motivate the relevant issues. I do each of these in turn.[24] First, what must be understood about moral responsibility in order to get the argument going has already been laid out in previous portions of this book; in particular the arguments about different views of how persons are bound up with other persons. As Matthew Talbert notes in his work surveying the topic of moral responsibility, three items are essential for understanding it. The first item is an emphasis on relationships and the commonality of moral responsibility practices. He writes that "[m]aking judgments about whether a person is morally responsible for their behavior, and holding others and ourselves responsible for actions and the consequences of actions, is a fundamental and familiar part of our moral practices and our interpersonal relationships."[25] Thus it should not be surprising that those involved in a physician–patient relationship (even if understandings of that relationship have changed over time) think of the other party to

[24] Much of this framing relies on Talbert (2024). Though an encyclopedia entry this is one of the more clear and more inclusive assessments of moral responsibility.

[25] Talbert (2024).

the agreement as exhibiting behaviors that are the kinds of things (e.g., their actions (or omissions) and the consequences of those actions (or omissions)) for which they can be held responsible. When a physician misses key markers indicating a disease and so misdiagnoses a patient, and thus treats them incorrectly resulting in negative health outcomes, the patient holds the physician responsible. They might ask why the physician missed what they missed? They might be upset or annoyed or resent the physician (emotions which may or may not be exhibited if the physician did everything right, but a similar outcome obtained). Though, of this dyad, physicians are often focused on for responsibility appraisals, they might think similarly of patients and the language is also similar. A patient in need of, for example, significant weight loss in order to successfully undergo surgery might be held accountable for failing to follow a prescribed diet. A patient who is in need of modified activities so as to preserve the stability of a joint or allow an organ time for healing, might be held responsible for modifying activities or resting.

The reason that both parties might think of the other's actions (and omissions) and the consequences thereof as the kinds of things that they can (and sometimes should) hold the other responsible for is, as Talbert notes, because of the power that each holds. This is the second important item. He writes, "The judgment that a person is morally responsible for their behavior involves…attributing certain powers and capacities to that person, and viewing their behavior as arising, in the right way, from the fact that the person has, and has exercised, these powers and capacities."[26] The physician is in possession of the powers and capacities to diagnose, and fails. The patient has the powers and capacities needed to follow a particular diet or to get some rest. When others (physicians or patients) fail or succeed in light of these powers and capacities, we (the other party) often praise or blame them, again highlighting the role of moral responsibility. This is because, "To hold someone responsible involves…responding to that person in ways that are made appropriate by the judgment that they are morally responsible. These responses often constitute instances of moral praise or moral blame…Blame is a response that may follow on the judgment that a person is morally responsible for behavior that is wrong or bad, and praise is a response that may follow on the judgment that a person is morally responsible for behavior that is right or good."[27] And ascriptions of praise and blame (as well as the possibility, noted by Talbert but which I do not give space to here, of being morally responsible for the obtaining of a state of affairs which is neither praiseworthy of blameworthy) and the third item.

Each of these items fits the well as part of the description of interactions in medicine: first, they are relationship-focused; second the relevant parties have capacities and powers that are relevant to, third, what they do or fail to do and, when a state of affairs obtains as a result of their so doing or so failing, the other party often praises or blames them, that is, they hold them responsible.

A final point to clarify what is meant by moral responsibility and to help to distinguish it from other concepts, before addressing conceptions of moral responsibility

[26] Talbert (2024).

[27] Talbert (2024).

and how this is the right point of focus, is needed. Moral responsibility is neither a specification of what physicians or patients are obliged to do nor is it merely an ascription of causal responsibility. Consider the aforementioned cases. It is true that in the contexts in which the situations likely occur, the physician may have an obligation to diagnose well; it is also true that the patient may have an obligation to follow a diet or to rest, though those are less clear in light of the ways that we often understand patient-centered care in medicine and the value-laden nature of dominant approaches in subfields of bioethics, such as clinical ethics. These cases thus serve as a useful demonstration of these points. Let us take them each in turn. First, if we are focusing on moral responsibility, we are "concerned not with accounts that specify people's responsibilities in the sense of obligations, but rather with accounts of whether a person bears the right relation to their actions to be properly held accountable for them."[28] The physician who fails to diagnose an obvious cancerous legion (obvious to persons of their training, skill set, professed practice, et cetera) may be morally responsible for the subsequent death of a patient if that patient eventually succumbs to that cancer when, if it had been caught at the time the physician viewed the legion, the patient might have avoided death by that cancer. However, suppose, if walking in town, a physician looks up and notices a runner pass them, passing close enough that the physician views a cancerous legion on the runner's exposed arm. Suppose also that that runner dies, in the same way that the patient in the first instance did. In this second instance, the physician would not (at least not on most accounts) be held morally responsible for the death of that runner, even—let us suppose—the physician being a runner themselves could have easily caught up to the runner and shared the news that they ought to have their arm checked out. The physician in the second instance might be causally responsible, but not morally responsible. According to Talbert, "We may assign causal responsibility to someone for an outcome that they have caused, and we may also judge the person morally responsible for having caused the outcome. But the powers and capacities that are required for moral responsibility are not identical with an agent's causal powers, so we cannot always infer moral responsibility from an assignment of causal responsibility."[29] The physician in the second instance could have prevented the death of the runner, and, one might say was in the causal chain, but did not bear the right relation to their actions—they had no expectation that they would have an opportunity to diagnose or to run that day, and had not prepared well for either, or the like. Consider also the previous discussion of the patient and of values and patient centered care. The aforementioned patients who failed to follow the necessary diet or who failed to rest might be causally responsible for the negative health outcomes, but given other factors that may make it extremely challenging or near impossible to do so, it might be ill-fitting to describe them as morally responsible; that is, we would not blame them for the negative health outcome.[30]

---

[28] Talbert (2024).

[29] Talbert (2024).

[30] Given the patient-centered nature of medical care and the values focus in clinical ethics, it may be that a different patient in a similar context possesses the powers and capacities necessary for moral

There is a great deal that could or might be said about moral responsibility. The topic has received significant treatment throughout the history of philosophy. Connections between moral responsibility and free will could be commented on: is compatibilism viable and does it matter in thinking about medicine? Must a physician possess the ability to do otherwise in order to be free? In order to be held responsible? Is responsibility more about emotions than these questions suggest? Is a reactive attitude account, describing how patients feel in response to errors, or a response to positive outcomes and tracking the ethically questionable "grateful patient" programs? Given the host of factors, not just healthcare, which contribute to the health of patients, how should we think about moral luck and health, relative to moral responsibility? There is much that could be discussed, but oddly enough for a book that argues that moral responsibility is the key to understanding the debate over conscience in medicine, these questions—though both interesting and important—are outside its scope. The main positive claim of this book is that the debate is about conceptions of moral responsibility, about how professionals who care for others might understand themselves to be bound up with those others in community and what that might mean. It is not the aim of this book to offer a particular account of moral responsibility, arguing in favor of one or against another—that is an interesting and important, but a different project.

For the arguments of this book to be successful, the inclusion of moral responsibility as a consideration about which it is reasonable for professionals to have a view, relative to appeals to conscience, is needed; as long as some of the diverse set of accounts of moral responsibility are available to physicians in this way, the framing succeeds. Because of this, it is other accounts of conscience (and what the normative implications of those accounts for the inclusion of moral responsibility are) that need to be addressed. And so, in the next chapter, I engage popular views rooted in autonomy within the debate over conscience, but before addressing those, it is important to be clear about what is at stake in focusing on moral responsibility. To be clear, there are two sets of issues and two sets of arguments under consideration; one set is about particular accounts of moral responsibility—that, again, is the task of another book. However, the second set of issues is at the heart of this text: there exists a diversity of accounts of moral responsibility, any subset of which are reasonable for persons to hold, including medical professionals. This is what is really at issue in the debate over conscience: CDs beg important questions against medical professionals who hold particular accounts of moral responsibility. Those dissenters attribute disagreement to religious views, to failures to treat patients as vulnerable or to put them first, or to fail to understand what their profession is (e.g., that medical professionals are actually semi-public employees). These claims are mistaken and the focus on religion, as has been argued, is a red herring.

I turn, in the next chapter, to conscience-specific considerations of moral responsibility in relation to arguments (primarily those rooted in autonomy), before moving to a discussion of medical actions and moral responsibility in the fifth chapter.

---

responsibility, but is not held to have failed an obligation because the health goals of the patient are, in these contexts, determined by the patient.

References

Engelhardt, Hugo Tristram. 2000. *The Foundations of Christian Bioethics*. Exton: Swets & Zeitlinger Publishers.

Engelhardt, Hugo Tristram. 2017. *After God: Morality and Bioethics in a Secular Age*. Yonkers, New York: St Vladimir's Seminary Press.

Evans, John Hyde. 2014. The History and Future of Bioethics: A Sociological View. 1. Issued as paperback. Oxford: Oxford Univ. Pr.

Howard-Williams, Mara, Rieza H. Soelaeman, Leah S. Fischer, Russell McCord, Robin Davison, and Christopher Dunphy. 2022. Association Between State-Issued COVID-19 Vaccine Mandates and Vaccine Administration Rates in 12 US States and the District of Columbia. *JAMA Health Forum* 3 (10): e223810. https://doi.org/10.1001/jamahealthforum.2022.3810.

Jones-Nosacek, Cynthia. 2024. Vaccine Mandates: Weighing the Common Good vs Personal Conscience and Autonomy. *The Linacre Quarterly* 91 (2): 147–167. https://doi.org/10.1177/00243639231213515.

MacIntyre, Alasdair C. 2007. *After Virtue: A Study in Moral Theory*, 3rd ed. Notre Dame: University of Notre Dame Press.

MacIntyre, Alasdair C. 2006. *Three Rival Versions of Moral Enquiry: Encyclopaedia, Genealogy, and Tradition; Being Gifford Lectures Delivered in the University of Edinburgh in 1988*. Repr. Notre Dame, Ind: Univ. of Notre Dame Press

Pilkington, Bryan C. 2018. The Fiftieth Anniversary of Patient as Person: Paul Ramsey's Groundbreaking Approach to Christian Bioethics. *Christian Bioethics: Non-Ecumenical Studies in Medical Morality* 24 (2): 111–125. https://doi.org/10.1093/cb/cby003.

Pilkington, B. C. 2019. Patients, Practitioners, and Conscience: Why Religion Should Not Divide. *Ethics, Medicine and Public Health* 11: 30–35. https://doi.org/10.1016/j.jemep.2019.100419.

Rawls, John. 2005. *Political Liberalism*. Expanded ed. Columbia Classics in Philosophy. New York: Columbia University Press.

Scanlon, Thomas. 2000. *What We Owe to Each Other*. First Harvard University Press paperback edition. Cambridge, Massachusetts London, England: The Belknap Press of Harvard University Press.

Smith Iltis, Ana, Ed. 2010. *At the Roots of Christian Bioethics: Critical Essays on the Thought of H. Tristram Engelhardt, Jr.* Salem, MA: M & M Scrivener Pr

Stahl, Ronit Y., and Ezekiel J. Emanuel. 2017. Physicians, Not Conscripts—Conscientious Objection in Health Care. *New England Journal of Medicine* 376 (14): 1380–1385. https://doi.org/10.1056/NEJMsb1612472.

Talbert, Matthew. 2024. Moral Responsibility. In *The Stanford Encyclopedia of Philosophy*, edited by Edward N. Zalta and Uri Nodelman, Fall. Metaphysics Research Lab, Stanford University, 2024. https://plato.stanford.edu/archives/fall2024/entries/moral-responsibility/.

The Belmont Report. 2014. Ethical Principles and Guidelines for the Protection of Human Subjects of Research. *The Journal of the American College of Dentists* 81 (3): 4–13.

Chapter 4
Moral Responsibility: Why Autonomy Misses the Mark

In the previous chapter, I argued that religion need not and, in fact, ought not be the focus of conversations and debates around the appropriateness of conscientious practice. At this juncture, it may be helpful to take stock of key points in the preceding chapters as a way of tracking the argument thus far. In Chap. 1, I distinguished between two clusters of views that emerge in discussion of conscience in healthcare. The first group, CPs, find practitioner judgments, including their practical reasoning about the situations their patients bring to them and their own self-reflections on their place in caring for the health of those patients, to be important (potentially essential) to their practice of medicine. This group was contrasted with a cluster of views falling into the category termed conscience dissenters. The latter group, as opposed to deferring to the practical judgement of practitioners, places primacy in considerations of conscience within the health professions themselves (not individual professionals, but "the profession"), within laws, or within other cultural, social, financial, or other kinds of regulatory standards. The bulk of the first chapter, after laying out the tension between the two clusters of views, presents an argument—against current trends—for the first cluster.

Persons, including professionals within the same health profession will—and do!—disagree, and medicine is no different. Recognizing and resolving disagreement is the key to understanding the debate over conscience. And so, Chap. 2 focused on the challenge of disagreement. There, I first took up conceptions of professions, relying on an account of health professions from the sociological literature to undergird later discussions of disagreement and its resolution. Part of these later arguments involved clarifying philosophical approaches which are appealed to by conscience dissenters, but which misunderstand (or, at least, appear to offer unsupported interpretations of) the theoretical resources to which they appeal. With these arguments in place, questions about the tenor of arguments in the literature were raised; in particular, about what appears to be an unargued for assumption that the burden of proof—in situations where tensions arise relative to a claim of conscience—is always on the practitioner who invokes conscience.

© The Author(s), under exclusive license to Springer Nature Switzerland AG 2025

B. Pilkington, *The Medical Act: Conscientious Practice in a World of Dissention and Disagreement*, The International Library of Bioethics 114,
https://doi.org/10.1007/978-3-032-00048-4_4

These arguments led to work in the third chapter on religious beliefs. There it was argued that the issues at the heart of the debate over conscience in medicine are not—contrary most positions in the literature and (anecdotally) the expressed views of most persons in conversations about healthcare practice—religious; rather they are about moral responsibility. It is to that topic to which I now turn in this fourth chapter.

4.1 What Things Look Like if We Take Moral Responsibility Seriously

I have argued throughout this book, especially in the final section of Chap. 3, that understanding moral responsibility is the key to understanding debates over conscience. This broader line of argumentation continues throughout the remainder of the text. The argument in this fourth chapter is primarily negative. In Sect. 4.2, I take care to explicate conceptions of autonomy operative in the debates over conscience and the implications thereof, including the language of values, concerns about values impositions, and commitments of autonomy-based positions. In so doing, I highlight three stages of recent thought on the role of conscience in medicine. These are not to be understood as historical stages; that is, they are not stages in time. Rather, they are stages in a dialectic, with amendments introduced and qualifications added to earlier positions that have led to distinct accounts within the cluster of autonomy-based positions. I have chosen representative samples of these positions, given their prominence, focus, and an ability to capture common sentiments about the relevant issues. In Sect. 4.3, I caution against using the description of conscientious objection in discussing a number of common cases that are better described as instances of conscientious practice, reaffirming and supporting claims made earlier in the text. Throughout this book, I have relied on the phrase "conscientious practice," as well as terms like "conscience protector" and "conscience dissenter." These terms, especially the latter two, bring with them their own rhetorical force and connotations. These terms carry with them potential linguistic bias, but as most terms in this area of the debate suggest some of the normative commitments that their proponents possess, flagging the language here is a sufficient disclaimer for the reader. Relatedly, the reminder helps to underscore that the critique of conscientious objection, as inapplicable to many common cases, is not merely an instance of preference. Conscientious objection simply does not fit some of these cases. This leads to a need for a positive response to the issues surrounding the role of conscience in medicine. In Sect. 4.4, the work of the ethicist Alasdair MacIntyre is engaged as a means of critiquing some contemporary understandings of bioethics, the arena in which many of the debates over conscience in medicine occur. This chapter concludes with an appeal for a public philosophical discourse. At this juncture, it will be beneficial to continue the sketch, albeit still briefly and inadequately, of what is meant by moral responsibility and why it matters to conscience and the debates over it in medicine.

For the purposes of this argument, and this text, I leave open what the particular account of moral responsibility ought to be. This is intentional. The points of interest are fourfold: First, persons view themselves and their relationships and interactions with others in different ways. Second, this is an important feature of persons for a number of different reasons and can be supported from a number of different philosophical positions (as well as rejected from a number of others). Third, to reject a person's conception of moral responsibility is a weighty thing, in light of the first and second points, and so it must be done carefully and in a way that attends to particular arguments, histories, and contexts—that is, the conception (and the person) must be taken seriously. Fourth, the resources required for this kind of engagement are philosophical, in a broad sense. I briefly take up each point, before moving to critically assess autonomy-based accounts, but first I explain why this task is appropriate (that is, why I intentionally leave open accounts of moral responsibility) and how the tenor of the arguments in the chapter might be understood.

4.1.1 Philosophical Engagement

The Lindley Lecture at the University of Kansas is given in honor of Ernest Lindley, who was the Chancellor of that university from 1920 to 1939. Those details could not matter less, except for the fact that he happened to be interested in and taught courses on topics like *living* and *the human situation*, which led to the charge that the invited lecturer must speak on "Values of Living." In 1984, the lecture was given by the then professor of philosophy at Vanderbilt, Alasdair MacIntyre, who asked a simple question, "Is Patriotism a Virtue?" It is a question worth taking up again, given both the current socio-political climate and the disagreement that pervades our public moral discourse and vocabulary, but also because we find ourselves where we do and that matters (and matters more than some of us are willing to admit). This text is not an analysis of virtues nor of patriots. However, the arguments in this chapter—and to foreshadow some of the benefits of the soon-to-be-offered solution—are, if successful, of the kind that MacIntyre describes in 1984. He opened the Lindley lecture in this way:

> One of the central tasks of the moral philosopher is to articulate the convictions of the society in which he or she lives so that these convictions may become available for rational scrutiny. This task is all the more urgent when a variety of conflicting and incompatible beliefs are held within one and the same community, either by rival groups who differ on key moral questions or by one and the same set of individuals who find within themselves competing moral allegiances. In either of these types of case the first task of the moral philosopher is to render explicit what is at issue in the various disagreements and it is a task of this kind that I have set myself in this lecture.[1]

[1] MacIntyre (1984).

This is a primary aim of this chapter and why a public philosophical discourse, one that brings the tools of philosophy to the public arena in order to resolve these kinds of disagreements, will prove useful.

Different persons engage with others, both individually and in community, and do so in different ways in light of their commitments, the histories of their lives and attachments, and the contexts in which they find themselves. This is borne out though a number of philosophical treatments of a variety of different issues, and done so via distinct approaches. It can be illustrated through debates over atomism and communitarianism in ethics,[2] though social and political commentaries about individuals who view themselves to be on the margins of society,[3] through legal scholarship on the status of persons and their individual rights as contrasted with rights that might apply to groups,[4] and through the oft referenced global and multidisciplinary work that aims to codify the grounding and implications of how we interact with others: the declaration of human rights. From this final example, the very first article of the document mentioned states: "All human beings are born free and equal in dignity and rights. They are endowed with reason and conscience and should act towards one another in a spirit of brotherhood."[5] Thus, a connection between conscience and how we interact with others, that is, how we are connected with them or how we might be bound up with them in human community is recognized, at least in this most influential international document.

Reflection on this connection leads to an understanding of the importance of respecting accounts of moral responsibility. It is also, working in reverse order, a key question in legal scholarship. The work of the legal scholar Jeremy Waldron illustrates this point well. Waldron's work on dignity is impactful and far-reaching. It includes histories of the concept, how it engages the status of persons, and—most importantly for our purposes here—the question of whether dignity (and rights dependent on it) ought to apply only to individuals or ought to apply to groups.[6] Though this is rarely the case, the answer to the question does not matter. Asking the questions is the crucial step. For it is how persons interact, how they perceive what their relationships with others might mean for their obligations and commitments to those others and the importance of such obligations and commitments to them, that the questions are raised at a foundational level in legal scholarship. Further, it is not only in prominent places in international human rights documents nor legal scholarship that these questions arise. One way to interpret the literary work of the writer and social critic, James Baldwin, is to ask how it is we are bound up with others in community and what might persons (knowingly, unknowingly, or merely in light of their location) be responsible for, not in terms of their accountability, but

[2] See, for examples, Taylor (1979), MacIntyre (1988) and MacIntyre (2007).

[3] For one helpful example of this from a perspective based in the United States, see: Baldwin (1993).

[4] See, for example, Waldron (2008).

[5] United. "Universal Declaration of Human Rights." United Nations. United Nations. Accessed July 2, 2024. https://www.un.org/en/about-us/universal-declaration-of-human-rights.

[6] See, for instance, Waldron (2015), as well as Waldron (2012), the latter of which was originally given under the title: Dignity and Defamation: The Visibility of Hate.

in terms of the prior question, how they are connected to others. Baldwin wrote, in *My Dungeon Shook: Letter to my nephew on the One Hundredth Anniversary of Emancipation*, 1962[7]:

> I know what the world has done to my brother and how narrowly he has survived it and I know, which is much worse, and this is the crime of which I accuse my country and my countrymen and for which neither I nor time nor history will ever forgive them, that they have destroyed and are destroying hundreds of thousands of lives and do not know it and do not want to know it. One can be–indeed, one must strive to become–tough and philosophical concerning destruction and death, for this is what most of mankind has been best at since we have heard of war; remember, I said most of mankind, but it is not permissible that the authors of devastation should also be innocent. It is the innocence which constitutes the crime.

What is both striking and fitting for our discussion of the role of connection to others, of how one might be bound up with others, is that Baldwin's assessment of the racist treatment he describes is about moral responsibility. Yes, it is also about accountability, but that is a later point, given—first—those claims of moral responsibility. What makes the destruction of lives even worse is that it is perpetrated unknowingly and by those who refuse to know what they do or of what they are a part. This is a claim of *moral responsibility*, and one could, with more careful mining of Baldwin's text, extract the necessary and sufficient conditions for ascertaining blame for intentional instances of and unintentional benefits from structural racism. The washing of hands, the avoiding of blame, the failure to take responsibility—these, borrowing from Baldwin—are the crime. This is a description of a system, one which can be assessed best through that key interpretive lens: moral responsibility. Later, in that same text, Baldwin claims that freedom is actually not something that many want, noting, " Freedom is hard to bear…Privately, we cannot stand our lives and dare not examine them; domestically, we take no responsibility for (and no pride in) what goes on in our country; and, internationally, for many millions of people, we are an unmitigated disaster."[8] Again, we see the emphasis on moral responsibility rooted in the communities, smaller and larger, of which we are a part.

Lastly, debates over the obligations of persons rooted in whether they exist as individuals as opposed to members of communities also highlight the importance of this general idea of moral responsibility. Before describing this last example, it should be reemphasized that what is needed for the argument of this text to be successful is that conceptions of moral responsibility—whatever those particular accounts are—are important enough to persons that they count as a part of their lives, of their moral lives, of their self-conceptions, or—to borrow language from contemporary bioethics—of their values. If this is the case, then the CD has quite a task in arguing against respect for those values and an important feature of a person's identity. This is illustrated well in the work of an ethicist who, though not a communitarian, places deep emphasis on community and, like Baldwin, recognizes the force of obligations, even unchosen or unnoticed ones. MacIntyre writes, rejecting

[7] Baldwin (1993, p. 5).

[8] Baldwin (1993, p. 87).

atomism and highlighting what might make persons responsible, including unchosen features of their social identities:

> For I am never able to seek for the good or exercise the virtues only qua individual…But it is not just that different individuals live in different social circumstances; it is also that we all approach our own circumstances as bearers of a particular social identity. I am someone's son or daughter, someone else's cousin or uncle; I am a citizen of this or that city, a member of this or that guild or profession; I belong to this clan, that tribe, this nation. Hence what is good for me has to be the good for one who inhabits these roles. As such, I inherit from the past of my family, my city, my tribe, my nation, a variety of debts, inheritances, rightful expectations and obligations. These constitute the given of my life, my moral starting point. This is in part what gives my life its own moral particularity.[9]

Our actions are to be governed by obligations which arise, at least in part, from our social identities. I owe things to my daughters that I do not owe to others in light of the fact that I am their father, just as my neighbors may rightfully expect certain kinds of behaviors from me which those who do not live in close proximity may not. MacIntyre's critique of individualism may be persuasive, but the point here is to highlight how this is yet another argument for how we might be bound up with others. In the background of claims of obligation, of accountability, of views of community are accounts of moral responsibility—sometimes explicit, sometimes implicit, but always present. MacIntyre continues his critique of individualism, capturing an idea that was explored in chapters two and three of this text, as well as later on in this fourth chapter: Of particular interest in this chapter is the idea that autonomy of professionals exists, but that it need not be respected in considering the autonomy of others or that it is fully respected merely because a person had an option to choose a particular profession; after that, no further respect for it is needed. Helpfully, MacIntyre writes, "This thought is likely to appear alien and even surprising from the standpoint of modern individualism. From the standpoint of individualism I am what I myself choose to be." And one might, as Stahl and Emanuel argue in a position discussed in the second chapter, simply give up being a healthcare professional and so respect for conscience is not needed. However, MacIntyre continues:

> I can always, if I wish to, put in question what are taken to be the merely contingent social features of my existence. I may biologically be my father's son; but I cannot be held responsible for what he did unless I choose implicitly or explicitly to assume such responsibility. I may legally be a citizen of a certain country; but I cannot be held responsible for what my country does or has done unless I choose implicitly or explicitly to assume such responsibility.[10]

This MacIntyrean thought gets to the heart of a deep disagreement about the role of conscience in medicine, and one framed in the terms that I employ in the argument of this text: moral responsibility. That disagreement is between those who find conceptions of moral responsibility to be part of the rich moral lives of persons and so believe that claims of conscience, which rely on a person's conception of moral responsibility, ought to be respected. Those who disagree with this claim appear to

---

[9] MacIntyre (2007, p. 221).

[10] MacIntyre, (2007, p. 221).

believe that one can choose, as MacIntyre notes, to assume or not assume particular responsibilities, that one can change one's conception of moral responsibility or that one cannot ask for it to be respected because it requires something of others.[11] Yet for MacIntyre, it is not only permissible to think of responsibility in this way, one cannot avoid these obligations, which—importantly—are obligations that are unchosen. He continues:

> Such individualism is expressed by those modern Americans who deny any responsibility for the effects of slavery upon black Americans, saying 'I never owned any slaves'....the Englishman who says, I never did any wrong to Ireland; why bring up that old history as though it had something to do with me? or the young German who believes that being born after 1945 means that what Nazis did to Jews has no moral relevance to his relationship to his Jewish contemporaries, exhibit the same attitude, that according to which the self is detachable from its social and historical roles and statuses.[12]

Persons are, for MacIntyre, story-telling animals who live (or seek to live) lives of narrative unity. If this is right, then we must attend to two normative implications of taking this description of persons seriously: first, in ethical considerations inside (and outside) of healthcare contexts, we must recognize that we need to know a person's story, but also what stories they are a part of, to make an ethically sound decision and to aid others in so deciding; second, if we take seriously the stories persons find themselves a part of, and they care about these stories, then we must respect their understanding of their own embeddedness and their own view of how they are connected to the other characters in these stories. We must defer to others regarding their accounts of responsibility.

To conclude this section, I remind the reader of what is needed for the argument to continue. This text does not offer an argument for a particular account of moral responsibility. Rather, it argues that conceptions of moral responsibility ought to count as part of a person's values (in the sense common to bioethics discussions) or part of their rich moral life. The diversity of accounts of moral responsibility—a survey of which is outside of the scope of this text—is not an obstacle to this argument, as all that is being claimed is that respect for such conceptions is needed. The position argued for in this text is not without costs. Though a focus on moral responsibility—as we will see later in this chapter—is explanatorily powerful, it also opens the door for abuse, division within communities, and raises yet another point about which disagreement might occur. These concerns are taken up in Chap. 5. I now turn to the aforementioned negative arguments and, first, to the failure of autonomy-based accounts.

---

[11] For more on this point, see the discussion of positions classified as falling into Stage Two later on in this chapter.

[12] MacIntyre (2007, p. 221).

4.2 Why Moral Responsibility Is Better Than Appeals to Autonomy

Encounters between physicians and patients can be some of the weightiest interactions in a person's life. Persons hold their health very dear, albeit among other things, and they often seek the care of physicians when their health is compromised or put in jeopardy. Approaching another person from a state of vulnerability—both in terms of a weakened or potentially weakened state but also the commonly vast difference in knowledge[13] regarding illness—is not easy and places a burden on the physician to take particular care in responding.[14] One feature of such care, incumbent on physicians to reflect on, is the autonomy-paternalism spectrum.[15] However, reflection on this spectrum and deference to patient autonomy, in light of—to borrow, again, the common language from bioethics—their values, need not mean that appeal to conscience by practitioners are inadmissible in care contexts.

This is not to discount the challenge that practitioners face. Physicians, and other healthcare professionals, are in the challenging position of aiding others without possessing complete information or full control. Though the former can be remedied to varying extents depending on the situation, the latter should not be. Physicians are not merely technicians, as the ethicist William May noted in a famous treatment of the role and images of physicians. Surely "[e]xcellent technical performance marks the professional;"[16] but, health professionals are persons who practice *with* patients, as members of a broader profession, within multiple traditions of practice, and often focusing on one of a variety of specialties. Were physicians technicians, then taking care to respect the autonomy of their patients and guarding against overly paternalistic approaches to their practice *with* a patient would be of less concern. Technicians merely execute, they do not have rich lives of their own or nuanced self-conceptions, let alone existing as mere sets of values. This is not to discount the lives of patients. Patients, as persons, possess values, goals, and interests that might be very different from the physician with whom they share in the therapeutic endeavor. They have their own life history, relationships, understandings of the world and their place in it, and reasons that they have sought out the care of a physician. Because the patient is a person and not merely a problem to solve, even if perfect technique is displayed, the physician has not satisfied the ethical requirements of the profession. Physicians ought to, even if it is challenging, engage (and attempt to place themselves in) other perspectives. They need to remember that though medical practice brings

---

[13] Put to the side first-hand knowledge of an experiential sort.

[14] This claim is not in conflict with the claim from Chap. 3, that the burden of proof is not on the practitioner who appeals to conscience. One might have a number of burdens in this space, some applicable and others not.

[15] This image is a rough approximation; a better, but less clear image is that of actions, behaviors, or approaches to the physician–patient encounter as gravitating toward different poles, an autonomy-affirming pole and a paternalism-embodying pole.

[16] See ethicist, William May's classic medical ethics text (May 2000) for a discussion of technicians and other images of physicians that guide ethical approaches to medicine.

great goods to many, ethical violations within its history strengthen the need for respect for autonomy in their current practice,[17] as does the overall change in context, helpfully summarized by Kilbride and Joffe:

> The rejection of medical paternalism in favor of respect for patient autonomy transformed the patient-physician relationship. Historically, medicine and society subscribed to the ethical norm that the physician's main duty was to promote the patient's welfare, even at the expense of the latter's autonomy. A central assumption of the paternalistic framework was that physicians, because of their medical expertise, knew best what was in the best interest of patients.[18]

This historical change in orientation *from* physicians, as those with knowledge of what is in a patient's best interest, informing patients what the course of treatment would be *to* a relationship of shared decision-making that has been described by terms like therapeutic alliance or "joint-adventurers"[19] is due, in part, to an emphasis on the concept of autonomy. This, in part, is what makes autonomy a likely—though failed—candidate for aiding in analyses of conscience. In fact, autonomy is one of the famous "four principles" of bioethics[20] and, some argue, the most important of the four.[21] The concept's role in ethics has its roots in philosophy, especially within the Kantian Tradition,[22] and focus on the concept highlights that persons are self-governing or, according to the Greek roots of the term, that they give the law unto themselves. Such persons are described in the Belmont Report, an early document codifying medical practice and research norms, as those "...capable of deliberation about personal goals and of acting under the direction of such deliberation."[23] It should be clear that a physician–patient relationship defined by the physician who knows best and who merely informs the patient about what course of treatment is to be taken would not be an example of respecting the patient's autonomy. In such an instance, the patient would not have an opportunity for self-rule. However, this is not what is required by a view of conscience that focuses on moral responsibility, nor does deference to autonomy in this way mean that it is the concept through which considerations of conscience must be analyzed. It is true that, following a Kantian-inspired

[17] Some historical examples include: violations of freedom in favor of "best interests" (for a discussion of the famous Dax Cowart case, see Engelhardt 1989), racism in medical research and practice (including, failures to attend to internalized racism (see Smith 2019), racism in research and race-based medicine (Brandt 1978; Johnson 2019).

[18] Kilbride and Joffe (2018).

[19] Ramsey (1970).

[20] Beauchamp and Childress (2013). Though principlism is only one approach to ethical medical practice, it is a dominant approach.

[21] Farber Post and Blustein (2015).

[22] See Kant (1785). However, it should be noted that Kant and scholars in this tradition may raise questions about whether common application of the concept within clinical practice follows the best interpretations of Kant. That interpretive discussion is beyond the scope of this treatment of the autonomy.

[23] National Commission for the Protection of Human Subjects of Biomedical and Behavioral Research (1979). It is worth noting that "early" here refers to "bioethics," not the thinking about these issues broadly.

interpretation of autonomy, a hallmark of autonomy-affirming practices would be the offering of reasons to persons to evaluate for themselves and then the opportunity to act on those reasons. Taking this seriously suggests a bidirectionality and highlights why mere appeals to autonomy in the conscience space are insufficient—both patients and physicians act!

In the context of the physician–patient relationship, respect for patient autonomy is exemplified through the informed consent process[24]; in fact, the Belmont Report[25] translates the principle of autonomy (which it refers to as "respect for persons") to the application of informed consent. A physician–patient relationship marked by physicians merely informing a patient what is in the patient's best interest is often described as paternalism. Paternalistic practices violate the autonomy of patients and illustrate a failure to treat them as persons. However, paternalism and claims of conscience are distinct. Though it is possible that physicians both seek to practice conscientiously and to act paternalistically, it would be a mistake to conflate requests for conscience protection with paternalistic approaches to medicine. Why this is the case will become clearer later on in this and the subsequent chapter. Though paternalism is to be avoided in the ethical practice of medicine, this is not to say that good physicians do not exemplify some virtues also possessed by good parents. Virtues of parents are the images at the linguistic roots of paternalism, such as care, compassion, and even great effort to safeguard. In highlighting the importance of respecting patient autonomy in medicine, my aim is to highlight the ways in which autonomy is to be supported—which are separate from considerations of conscience. Again, as May notes, "the healer overreaches when he or she justifies overriding the patient's wants, wishes, decisions, and judgment on the grounds that the adult patient is a child, incapable of knowing his or her own good."[26] This kind of claim, a claim of what is better or worse for the patient is different from a claim about protecting one's conscience or a request for one's account of moral responsibility to be tolerated.

Paternalistic relationships suggest exactly what May describes, that the "parent" figure—in this case, the (paternalistic) physician—knows better. Such models for physician–patient relationships often fail, ethically, as noted above, because even if technical experts in their craft, physicians are not rulers over their patients' goals and aims. As highlighted in the recent medical ethics text on the Hippocratic Oath, "A technique, in itself, does not include determination toward an end and away from what opposes that end while an ethic necessarily does."[27] What distinguishes patient autonomy claims from practitioner claims of conscience and why respect for patient autonomy need not mean getting rid of conscientious practice or the protection thereof is that the ends and aims are in distinct conceptual spaces. Good collaboration between patients and conscientiously practicing physicians is still possible; in fact,

---

[24] This is especially the case when informed consent is not thought of as a document to be signed but as part of a process of shared decision-making.

[25] Department of Health, Education, and Welfare and National Commission for the Protection of Human Subjects of Biomedical and Behavioral Research (2014).

[26] May (2000).

[27] Cavanaugh (2018, p. 141).

with appropriate promulgation of views and values, it is likely. Shared decision-making models are most fruitful because they bring together the expertise of both the physician and the patient.

It is important to recall that respecting the autonomy of patients is rooted in the recognition that they are (or are capable of) self-governing and one might be tempted to note something similar of physicians and so move to describe appeals to conscience as a disagreement between the autonomy of a professional and the autonomy of a patient. This framing is commonly problematic for two reasons. First, in some instances, it conflates different conceptions of autonomy. Second, in others instances, it imports other concepts to fill in the content for autonomy, bringing in features which are in need of justification, but due to the manner in which these features are introduced, justification is not offered. That said, the rhetorical power of these kinds of claims of reciprocity should not go unnoticed. For instance, even if a patient shares a good deal of information about her life history with her physician, the physician would still not possess sufficient information to be justified in acting paternalistically, that is, to make decisions about the patient's care without the patient for the good of the patient. Even if such decision-making were ethically acceptable, a physician cannot fully embody the perspective of her patient. A good physician will learn enough about a patient to be empathetic, but not enough to decide for her. Something similar might be said about professional colleagues or institutional representatives vis-à-vis an individual practitioner's claim of conscience. Though this is not an argument offered in this text, if this is what motivates those who see this disagreement as one of autonomy, the impetus is understandable.

To conclude this section, one might summarize it as follows: an ethical approach to medical encounters requires that physicians see their patients as persons. Failure to do so, not only by physicians but by the healthcare and societal institutions of which they are (and have been historically) a part, has led to (and can lead to further) unethical treatment of patients. One feature of treating patients as persons is respecting them as autonomous agents, but this does not mean that claims of conscience must be framed as one person's autonomy versus another's, nor does it mean that conscientious practitioners cannot both appeal to their conscience and respect the autonomy of patients. It is to autonomy-focused accounts of conscience that I now turn.

4.2.1 Autonomy-Based Accounts of Conscience

One might accept the arguments in Chap. 3, or at least the conclusions of those arguments, and believe that considerations of conscience are not (or, at least) need not be about religion or even about personal philosophical views or commitments. Yet, one might not accept the view that what is at the core of this disagreement is different accounts of moral responsibility and how persons interact with others. One popular alternative focuses on the concept of autonomy. The previous section is a helpful illustration of why this might be the case. However, in so doing, these approaches pit autonomy of patients against the autonomy of professionals and seek to resolve this agreement through appeals to the concept of autonomy and the normative implications of reflection on it for healthcare practice. In this section, I take up these accounts, dividing the dialectic into three stages of this form of argument with the aim of illustrating how and why they are found wanting.

Before arguing against an autonomy framework for discussions of conscience in healthcare, it is important to highlight why, in addition to what might be gleaned from the previous section, these views are attractive. They are attractive for at least three reasons. First, autonomy is an important, influential, but also popular concept with bioethics, the field in which much of the discussion of conscience as it relates to healthcare takes place. Autonomy is, at least in the way that it is sometimes applied in clinical interactions, a more straightforward and simple account (than it actually is conceptually and should be practically) wherein respecting patient autonomy means that practitioners simply defer to patient choices. This is one of the reasons it is popular; it is clear and easy to apply. Second, framing the debate over conscience as a tension between appeals to autonomy from different parties (practitioners and patients) is conceptually cleaner than taking seriously a variety of conceptual resources that might bear on the discussion and then attempting the very challenging (and very real) task of trying to adjudicate between these resources. One move, as we saw in the previous two chapters (and especially in Chap. 2) is to focus on a location for resolution (for example, a professional association's meeting). Once the conversation is so restricted, and the available conceptual resources so limited, a resolution may appear possible. In the type of case mentioned, common responses to yield an appropriate outcome are proceduralist, further limiting the options. Finally, there are historical connections between invocations of autonomy and considerations of freedom. As the concept (and practice) of conscientious objection is often connected with the rights of persons to express sincerely held beliefs and for their integrity not to be violated; for example, the rejection by pacifists to fight in a war, it is not surprising that autonomy is the preferred concept of many to help understand and clarify issues in this space. I take up each of these motivations for the employment of autonomy, in turn, via a three-stage discussion.

4.2.2 The First Stage

The aforementioned idea that patients—and practitioners—are persons is important for the argument in this chapter and in the overall text. This becomes especially apparent in critiques of the framing of issues related to conscience in medicine as tensions between competing claims of autonomy. Before critiquing these views and positing an alternative, descriptions of the positions are needed.

There are a variety of ways in which one might track the history, or tell the story, of this strand of arguments about conscience. There will undoubtedly be different descriptions focusing on distinct interlocutors and other contributions to the intellectual conversation. The points offered here are appropriately illustrative of the relevant issues in autonomy-focused accounts and exchanges related to them. The narrative begins with a common ground position, one which seeks to find agreement (or a workaround) between two competing claims: what are sometimes called traditional claims of conscience and what are sometimes called contemporary claims of conscience.[28] The aim of this approach, which is well argued for in an important article by Mark Aulisio and Kavita Arora, now written almost a decade ago, is a compromise and a good one. Aulisio and Arora seek to respect the conscientious practice of individual practitioners and the autonomy of patients by distinguishing between what they view as two different kinds of conscience claims. They argue that traditional claims of conscience ought to be respected, but that contemporary claims ought not be. Given that both these sets of claims are quite different, they note, "This difference in kind is profoundly significant; so much so, we contend, that it puts them at odds with the normative basis for protecting conscience claims in United States health care settings in the first place, making them illegitimate."[29] Thus, for Aulisio and Arora these categories are importantly distinct, and a great deal hangs on the distinction, as claims in one category are legitimate and in the other, they are not.

By traditional claims of conscience, which they argue ought to be protected, they have in mind objections to the performance of certain actions which might be deemed by the performer to be—per their article's pithy title—evil. There is a long history of debates and arguments around complicity in or participation in evil in certain moral theories[30] and in certain bioethical reflections on healthcare practice,[31] but retelling that history is outside of the scope of this text. Rather, what is important for the purposes of this argument is how Aulisio and Arora, as a representative instance of a first view of autonomy-grounded conscientious practice, describe what ought to be protected. They write:

> The last 10–15 years of popular discussion surrounding conscience issues in health care saw a surprising reemergence of conscience claims following a period of relative quiet in the 1980s and 1990s. Most discussions of conscience in healthcare previously involved objections to direct participation in either abortion or what some termed "passive euthanasia" in the form of withdrawal of life support or "letting die." These objections were handled through a practical consensus of informing patients of their medical options and then arranging for transfer of care, referral, or alternate staffing as necessary (May 2002). At the institutional level, medical services that were deemed to be at odds with the core values of the institution were simply not offered by the institution (e.g., tubal ligation or in vitro fertilization at Catholic hospitals). We will refer to these types of claims as *traditional* claims of conscience.[32]

The reason for quoting their description at length is twofold. First, it is a helpful description of their understanding of claims of conscience as connected to a particular period of time in which they believe consensus held and where practical solutions were possible: inform patients and transfer them or refer them to another practitioner,

---

[28] Aulisio and Arora (2014).

[29] Aulisio and Arora (2014, p. 254).

[30] The work of Thomas Aquinas is influential on the subject. For a recent treatment of Aquinas on complicity, see: Mellema (2016) or Flannery (2001).

[31] See, for example, Flannery (2012) or Pilkington (2014).

[32] Aulisio and Arora (2014). Note, the internal reference is to: May (2002). May also coauthored a piece on conscience and informed consent with Aulisio.

if one was unable—for reasons of conscience—to perform the requested or needed intervention. Second, it is helpful to understand to whom claims of conscience apply. Claims of conscience are—with a few recent, notable exceptions—almost always attributed to individual persons: physician *x* conscientiously objects to institutional offerings of MAID because of their deeply held beliefs about *y*, and so will not prescribe *z* medication, but will refer the patient to physician *w* for the express purposes of that prescription. It is interesting that Aulisio and Arora note, under the description of traditional claims of conscience, that some institutions simply did not offer medical services which were in line with their core values, which appears to fall within their category of traditional claims of conscience. Discussions of institutional conscience are briefly touched on in the fifth chapter of this book.

The second category of conscience claims, what Aulisio and Arora refer to as contemporary claims of conscience are not to be protected on their view; in fact, as noted above they are considered illegitimate. This is because contemporary claims of conscience are not requests for protection from performing actions that are understood (by the performer) to be evil, but rather would only *involve* the relevant agent *in* an evil. Avoiding such involvement, if not a direct action, is not to be protected in their view. They write, "If our analysis of what we have termed contemporary conscience claims is correct, all of those considered…are objections not to committing an "evil" act itself, but to being somehow *complicit* in an "evil" act that might be committed by others."[33]

For Aulisio and Arora, the first category ought to be protected; the second ought not. This is because of a kind of reciprocity or what is sometimes referred to as the consistency approach,[34] and it is a reciprocity rooted in respect for another's autonomy. They write, first:

> In healthcare settings, this emphasis on liberty is captured by the value of patient autonomy which normatively undergirds the centrality of the practice of informed consent (at least for adult patients with capacity), the right to refuse treatment, the development of advance directives, the role of surrogates in decision making, and an entire constellation of related patients' rights. Indeed, one way of thinking about the historical emergence of clinical ethics in the U.S. is that it was an outgrowth of the civil rights movement—a recognition that an individual's right to live by his or her values (autonomy) does not disappear simply because one becomes a patient …this is the foundational reason for the ironclad nature of the right to refuse treatment altogether—*in other words, the right of a capacitated adult patient not to have the values of health professionals (or others) imposed on them.*[35]

Next, they note:

> Interestingly, in our view, it is the very same right to live by one's values– the societal emphasis on individual autonomy—that is the normative basis for protecting conscience claims in healthcare (and elsewhere). Health professionals do not lose *their* right to live by their values simply because they have taken up the practice of medicine, nor do, by extension, healthcare institutions or organizations have to abandon the core values that inform their mission…The normative basis for protecting conscience claims in U.S. health care settings

[33] Aulisio and Arora (2014, p. 263).

[34] Pilkington (2016b).

[35] Aulisio and Arora (2014, p. 260).

> in the first place, then, *is to prevent values imposition;* specifically, an imposition of patient values on health professionals (and by extension health care institutions) at the expense of their deeply held moral (or religious) beliefs.[36]

Thus, the argument is framed in terms of autonomy and reciprocity (or consistency). The normative basis for conscience protection on their view is rooted in respect for the autonomy of persons, and persons do not lose this autonomy either by becoming patients or by becoming medical practitioners. But what is meant by autonomy? On their view, it is the right to live according to one's own values and, particularly important, to not have another's values imposed upon one. Two features of this characterization are worth noting. First, as previously mentioned and as we see reiterated here, conscience includes—for some interlocutors in the debate—institutions. Second, it is worth highlighting that the focus is—as with the framing of these issues that begins this chapter—on persons. It is persons who have these rights not to have values imposed upon them, even when they are in conditions like being a patient or a medical practitioner. This point will fall out of the autonomy-based argument, at later stages in the dialectic, which is important to note, as is the potential tension between accounts that allow in institutions and those that restrict the possession of conscience to persons.

In many ways Aulisio and Arora offer a commonsense argument, rooting it in an influential principle of bioethics and an important value in the society in which they live and work (and, surely, within many others). However, their account fails for two reasons. First, the distinction between traditional and contemporary claims of conscience does more work than their argument supports. There exists a plethora of cases in which one might reasonably think of oneself as responsible for or hold oneself accountable because of something that was done by someone else. I lend you a saw and, instead of doing some artistic woodworking, you proceed to cut down a neighbor's beloved Japanese maple tree. In such a case, I may feel badly about what occurred but I—on most accounts—should not be held responsible.[37] In the very least, the details of the case can be written in such a way that this is surely the case; for example, you told me that you planned to do some woodworking and needed a saw for an artistic rendering of a beautiful tree and, as it turns out, you have been relatively reliable in the past (what you have said you would do, you did do). However, consider a variation of this case wherein you complain to me about our shared neighbor and describe how you are planning to get back at her from some slight that offended you—stepping onto her property and cutting down her favorite tree—and you ask to borrow my saw. If I give you the saw in the second case, I ought to be held to blame—I knew what you were going to do and what I did aided you in your endeavors. But for what exactly am I responsible?

This book is not a work in action theory nor is it a book describing a new account of moral responsibility, so a full discussion of what I or you are blameworthy for in this case and why is beyond the scope of the text. However, this is not needed for the argument at this juncture. All that is needed is that it is plausible and even

[36] Aulisio and Arora (2014, p. 260).

[37] I may be causally responsible, but I would not be morally responsible.

reasonable for me, in this case, to hold myself morally responsible for my role in the cutting down of the tree. Suppose I do hold myself responsible—for whatever reason—what resources are available from the autonomy-based reciprocity account of conscience protection to distinguish these cases? There are none, which is the first reason why the argument fails—it is unclear why someone could not value how they are connected with others and with situations greater than themselves. Some persons may draw a distinction between the actions that they perform and those that they merely contribute to, but others may not draw the distinction in the same way. Others may hold an account of moral responsibility for which they are to be held blameworthy (or praiseworthy; one can, of course, be responsible for positive states of affairs, as well). If one does not draw that distinction in the way that Aulisio and Arora do, the power of their argument dissipates because the distinction collapses as they have given no argument sufficient to support it. By this it is not meant that we can no longer distinguish between performing an action and being complicit in one, rather it is meant that the normative basis of autonomy and its articulation as a right to avoid values imposition fails to support one category over another if persons value a particular account of moral responsibility or, to put it another way, if persons value their own understanding of how they are connected to others.

A second reason that this account fails has to do with a discrepancy that arises between a laudable point of the view—its focus on persons—and the kind of picture of persons that must be in play if that distinction, which was obliterated, were to hold. Persons must be thought of as lists of values. Were that the case, then one might be able to describe one value set bumping up against another value set, and adjudication could be possible as—on this simplistic view—the unit of measure is the same: values. However, human beings have and live much richer lives. Another way of putting this is to say that "Human lives, human understandings, schemes of human values, are richer than mere articulations of discrete and isolated lists of values."[38] This is problematic for the autonomy-based position under consideration because this description is one that they would (or, at least, should) want to adopt, given their claims about persons, but if they do, there is no reason to rule out conceptions of moral responsibility as part of the rich lives of human persons. Or, as described elsewhere, " If liberty is our focus, if autonomy is the foundation of conscience protection, why limit this to the impoverished account of our moral lives in terms of a list of values, especially given the importance of interactions between persons? The answer is that [this account] lacks the resources to make sense of a more complex moral life."[39] This is a particularly bad failing, given their focus on liberty and autonomy. According to their description of the former concepts, "'liberty' understood as a core constitutional value implies (1) that individuals (and communities) have an equal right to live by their values and (2) that each party should be given maximal play to live in accord with their values while avoiding values imposition; that is, either imposing their values on others or having the values of others imposed on them."[40] It is unclear how

[38] Pilkington (2016b, p. 5).

[39] Pilkington (2016b, p. 7).

[40] Aulisio and Arora (2014, p. 7).

on a view that aims for maximal play to live according to one's values, one could not hold one's own account of moral responsibility. To conclude discussion of this stage of autonomy-based conscience protection arguments, though reciprocity and autonomy are both valuable features of this position, the way it is constructed relies on an unrealistic view of persons and of their self-conceptions: we are not mere list of values—that is too simple, but once we take better account of the richness (and messiness) of human life, there is no good reason (at least no argument that has been offered) as to why different accounts of moral responsibility cannot be accepted or, at least, held to be true by persons and operative in how they seek to live their lives.

4.2.3 *The Second Stage*

A second stage in the autonomy-based conscience protection arguments is to offer something akin to what Aulisio and Arora contribute, but to build up—much more robustly—the account of autonomy as liberty with the hope of avoiding the objection that different accounts of moral responsibility must be admitted. Greenblum and Kasperbauer offer one such articulation. Like the work of Aulisio and Arora, their arguments and analysis have merit but, in the end, ultimately fail. Also, like Aulisio and Arora, they serve as a helpful representative of a cluster of common positions in this space and, in particular, this stage.

Greenblum and Kasperbauer seek to improve on Aulisio and Arora's argument by deepening the account of autonomy and by offering, what they believe, is a more realistic appraisal of the situations of conflict in which conscience is often appealed to. They note:

> Aulisio and Arora do not fully appreciate the value imposition placed on physicians and healthcare workers in these cases. Simply appealing to value imposition cannot explain why non-traditional claims of conscientious objection should be rejected. Because someone's—either the patient's or the physician's—values will be violated in these kinds of cases, value imposition cannot do the explanatory work that Aulisio and Arora need it to.[41]

It is important to notice two implications of this critique, implications that are hallmarks of this second stage of arguments around conscience. First, this realistic approach forecloses the possibility of reconciliation between the two parties or alleviation of the tension between them through some kind of compromise. In so doing, it moves away from considerations of conscience as germane to persons as no longer the unit of analysis; that is too broad. Rather, what is being looked at are patients and practitioners. In other words, we are no longer at the level of persons, or of human beings, but now at the level of embodied social roles. Second, it illustrates how a conception of moral responsibility might be beneficial, in recognizing that Aulisio and Arora cannot rule out values of one party or another and also that one's values might be bound up with or require acceptance of particular accounts of moral

[41] Greenblum and Kasperbauer (2018, p. 314).

responsibility. They rely on the first in order to support the distinction between traditional and contemporary claims of conscience. Now, let us consider the details of their account.

In offering their account of autonomy (*autonomy as liberty*), Greenblum and Kasperbauer take seriously the need for independence in order that the exercise of autonomy is worthwhile. They cite influential work in bioethics, such as the (four-principle) principlist approach of Beauchamp and Childress, as well as important work in political philosophy, such as John Rawls's theory of justice, in describing their account:

> we should look to the broader conception of autonomy that can be found in classic liberal thought. Although the right to live in accordance with one's values is unquestionably morally important, autonomy consists of much more than this. In particular, autonomy requires independence or the capacity to independently set goals for oneself, which practically requires a sufficient understanding of one's situation and a sufficient level of freedom from controlling influences.[42]

Due to this richer notion of autonomy, one will not be able to resolve the issue in the way that Aulisio and Arora attempt to do so because compromise is not possible. Greenblum and Kasperbauer offer a resolution to the disagreement, by siding with one party over the other: they simply state that values conflict will occur and in any such case, values conflicts should always be decided in favor of the patient. They write, "Patients…enter their physician's office with an understanding that the physician's professional judgement, not their privately held values, will determine the advice they receive. Thus, the value imposition on physicians is not only permissible but in some cases necessary for adequate care."[43]

A hallmark of approaches in this stage of the debate is twofold, as noted above. First, they move away from a conception of the person as relevant to discussion of conscience to a conception of a particular role that one embodies as relevant. Second, conflict is resolved simply be picking a side. For Greenblum and Kasperbauer, the patient always "wins" in these cases. How can such a view respect the autonomy of both parties? Another way of putting this question is to ask why this view ought to be placed in the autonomy-focused category at all. The answer is that, for Greenblum and Kasperbauer, "physicians exercise autonomy in choosing their profession and the institutions they work within. As part of their profession, it is understood that they must provide medical advice that might conflict with their personal values."[44] Thus it is still a view that respects the autonomy of both patient and practitioner. Notice, however, that it does so by moving away from accounts of autonomy that are more similar to each other (like those employed by Aulisio and Arora), and contrasts autonomy in choosing a profession with autonomy of a patient in a healthcare practice context. The argument holds together because, unlike Aulisio and Arora, Greenblum and Kasperbauer are not interested in persons, but in patients and practitioners. These

[42] Greenblum and Kasperbauer (2018, p. 315).

[43] Greenblum and Kasperbauer (2018, p. 315.

[44] Greenblum and Kasperbauer (2018, p. 315).

two separate social roles allow for the application of different norms, norms that could not apply to arguments in stage one and still be consistent.

Focusing on patients and practitioners, as opposed to persons, matters for at least two important reasons. First, it introduces (or, at least, emphasizes) different categories or different spheres of morality[45] that might apply to each participant in the physician–patient relationship. This potentially shifts the burden of proof to one member of this dyad, and this is exactly what is intended by the proponents of this view and proponents of similar views in this stage, such as that of Julian Savulescu.[46] Those arguing for positions that fall into stage two, conceive of a healthcare professional's role as prior to their existence as persons. This allows for a resolution to tensions that might arise between patients and physicians (or physicians and healthcare institutions vis-à-vis their relationships to patients[47]) relative to conscience by removing the possibility of a valid claim of conscience in light of a person's role in the relationship. These positions still ought to be understood as falling within stage two, as they continue to frame and address this issue as one of autonomy—even if they introduce two accounts of autonomy—one, the common notion in bioethics about control over one's care options and another, relative to choices one might have in a social, political, and cultural climate in which one might choose one's profession. Proponents of this type of view think of practitioners as public entities; a representative of this view describes physicians as "semi-public employees," highlighting three key features of the physician–patient relationship: first, that the patient is vulnerable, second that physicians—as just noted—are semi-public employees, and, third, that physicians must treat patients as social equals.[48] Before turning to four substantive critiques of positions in this second stage, let us briefly consider each of these three features, highlighting where there is common ground and agreement between different analyses in this space, as well as where stage two positions diverge from those in stage one and also from the position for which I argue. First, that patients are vulnerable and that their vulnerability needs to be taken seriously is, at least in current ethics conversations and practice, accepted—as it should be. The asymmetry of information, positional standing, and the vulnerabilities that can come with states of illness have all been well discussed. As argued above, respecting the conscientious practice of individuals need not mean ignoring the vulnerability of patients nor should conscientiously practicing physicians do so; on this first point there is agreement across positions in both stages and the position argued for in this text. On the third point, that patients ought to be treated as social equals, this is also an accepted and fairly commonplace expectation of health professionals, at least currently. As with taking seriously the vulnerability of patients, there is no disagreement between the positions under consideration on these points. The example offered

---

[45] For a nice treatment of how roles in healthcare relationships—in this case specific to physicians—may generate distinct set of norms, see: Doernberg and Truog (2023).

[46] Greenblum and Kasperbauer serve as representative examples of this position, but similar kinds of arguments have been offered by others, including: Savulescu (2006).

[47] Cummins (2021).

[48] Greenblum and Kasperbauer (2018, p. 316).

of a physician treating a patient as a social equal is that "it is impermissible for a physician to mock his patient's life plan or sense of identity, mistaken or confused as it might be."[49] Again, agreement exists about this case. This kind of humiliation and mistreatment is unethical and should not be accepted in medical practice.[50] It is on the aforementioned second point that stage two positions diverge from my own, as well as from stage one positions. The former argue that physicians are semi-public employees for two reasons: first, because physicians receive government funding; second, because physicians (on their account) are similar to public officials. The first point is well-taken, but it misses the mark. It is true that given compensation structures relative to physician's work within a variety of types of health institutions, government funding surely makes its way into most physicians' pockets. Yet, this point misses the mark because it paints with a broad brush physicians who might be compensated in different ways and to varying degrees. Consider physicians in private practice or working in a hospital that is not state run and contrast these physicians with those working in, for example, a Veteran's Affairs hospital. The latter set of physicians might be under different kinds of obligations and might fit more closely the description of semi-public employees. The persuasiveness of Greenblum and Kasperbauer's analysis does not translate to physicians in private practice, those who work in other kinds of health systems, or those who operate under a concierge medicine model and may not accept insurance.[51] A second reason that this point misses the mark is that it does not get to the heart of the issue. It is not the money trail which matters here. (In fact, were the trail to be traced all the way back—and not artificially terminated with current Medicare and Medicaid reimbursement prescriptions—the analysis might require a survey of those who pay taxes in the US, something that may very well yield a diversity—including some deeply problematic (and contrary) accounts of health and healthcare—of which Greenblum and Kasperbaur's account would not be able to make sense.) What is at the heart of the issue is the relationship between patients and those who care for them. It is to this stronger point that I now turn.

The potentially strongest point offered, in analogizing physicians to semi-public officials is that—contrary to the earlier claim that agreement exists about treating others as social equals—when physicians aim to practice conscientiously, it simply follows that they do not treat their patients as social equals. This is illustrated with a case of referrals. Greenblum and Kasperbauer write:

[49] Greenblum and Kasperbauer (2018, p. 316).

[50] One reason that Greenblum and Kasperbauer may take pains to make this point, even though it appears both obvious and a standard part of contemporary medical practice is because their work in rooted in the concept of autonomy, and an interpretation of that concept, *autonomy as liberty*. Certain violations of persons, such as humiliating them, are more challenging to explain the wrongness of on accounts of autonomy, as opposed to accounts rooted in other concepts. For a treatment of this issue and a rejection of humiliating practices rooted in the concept of dignity, see: Pilkington (2016a).

[51] This response, and the majority of the examples in this text, is rooted in a focus on healthcare practices in the US; and this response may not apply to healthcare practice is other locales where healthcare is structured differently.

> The analogy between physicians and public officials also explains why it is wrong for a physician to refuse to refer. When physicians refuse to refer on grounds of conscience, they substitute their judgement for their patient's. In so doing, they impermissibly harm the physician–patient relationship by treating their patient as a social inferior. Therefore, physicians incur a role-based obligation to rely solely on their medical opinion and publicly informed duties. They must refer.[52]

If it is true that to not refer a patient is to fail to treat them as a social equal, then a significant objection to an account of conscience rooted in moral responsibility has been raised. However, it is not true, and this is for at least two reasons. First, physicians may be uncomfortable referring or supporting a care request for a variety of reasons rooted in the way that they aim to conscientiously practice. They might "in good conscience" choose to have a hard conversation with patients and their families about a host of requested interventions that are deemed to not be beneficial or what used to be called "futile,"[53] for example. Greenblum and Kasperbauer must view such a justification as an instance of failing to treat another as a social equal because it is an example of substituting the physician's or the care team's judgement(s) for that of the patient. This, however, is not the case, as there are different spaces of judgment and to argue as if physicians and patients occupy the same is to conflate two social roles that were distinguished earlier in their argument. To flesh out this discussion, consider the common case of requests for CPR in persons who would not benefit from it and the less often mentioned follow ups of slow codes and show codes.[54] This is the strongest case for Greenblum and Kasperbauer's analysis, but it is rooted in one which relies on the best medical science and the expertise of the practicing healthcare professionals. Though there are surely cases where physicians have attempted to offer what they believe best and not what the patient might value, in this case a physician might in their best judgement deem the intervention as ill-fitting as it will not lead to the patient's (or family's) desired outcome. If this constitutes a problematic instance of substituted judgement, then physicians ought not to be thought of as semi-public employees but rather as salespersons of a health commodity or as purveyors of fine health goods, offering whatever a patient orders.

A second reason the claim fails to be true is that it ignores the kinds of practices of engaging with others and interpersonal interactions which are indicative of treating another as a social equal and can be employed by conscientiously practicing physicians. Consider a physician who is uncomfortable prescribing for MAID,[55] and suppose this physician is so uncomfortable with this practice that they will not refer specifically for MAID. Suppose further that such a physician made this known to patients, to the relevant members of their institution and colleagues, and did so prior to their employment. Suppose that the reasons motivating this physician were

[52] Greenblum and Kasperbauer (2018, p. 317).

[53] For a discussion of this kind of recommendation and the importance of honesty in medicine, see: Kaldjian and Pilkington (2021).

[54] For a justification of deceiving patients, one which remains controversial, see: Meyers (2021).

[55] Though an awkward phase, "prescribing for," is to capture what is being done while maintaining neutrality between those who describe the agents prescribed in these cases as "medications" and those who find that description problematic or objectionable.

primarily twofold. First, the physician sought to become a physician, interpreted an oath that they professed (at the beginning and conclusion of medical school, reaffirming it every time they attended a medical school commencement or whitecoat ceremony), and practiced their craft with an understanding of medicine as always to heal and they believe that MAID is not an instance of healing. Second, suppose they hold an account of moral responsibility wherein they believe that by prescribing for MAID, they are actively involved in the death of the patient, not in an act of healing, and that in referring a patient specifically for MAID they would be doing something similar. If in a shared decision-making conversation, the physician reaffirmed in a supportive, honest, and clear manner that they did not participate in this kind of care, it is not clear how it is—and, in fact, it simply appears false—that the physician is not treating the patient as a social equal. What this case illustrates is that respecting another as a social equal does not preclude disagreement nor does it require that one party must always to acquiesce to the other. Further, suppose that this physician did not attempt to dissuade the patient from accessing MAID services, they simply informed the patient—information that had been appropriately promulgated previously and which is known by colleagues and others—that they are not comfortable with the practice and that they do not participate in it. Again, it is unclear how the patient would be being treated as a social unequal but further, it is not clear how—on any account of substituted judgment—this physician is substituting their judgment for the patient. The only such account wherein this argument would work is if one defined substituted judgment as any situation in which a physician did not offer (or refer for) a service that was requested, and that definition both fails to capture what is usually meant by substituted judgment and also begs the question against even the possibility of conscientious practice.

To conclude discussion of stage two positions, I briefly highlight four critiques related to the notion of semi-public officials, to which this position opens itself up. Three of these critiques are interesting and relevant. One of these critiques, as argued elsewhere,[56] is decisive. The analogy between physicians and government employees, which much of the argument from Greenblum and Kasperbaur relies on, is ill-fitting, but further consideration of the implications of such a view is merited. Their account of autonomy as independence—that attempt to build up some of the conceptual resources offered in stage one arguments (in particular, of which Aulisio and Arora's work was the representative sample)—is a formidable one, given the vulnerable nature of patients and information asymmetries between patients and practitioners. In addition to their referral requirements, they claim that patients should be fully informed—something, again, about which there is common agreement—and this includes sharing information about rights of patients and the "the full range of legitimate medical options."[57] This is due to their conception of medical professionals as government employees and builds on (or, at least, is in line

[56] For an earlier version of this argument, see: Pilkington (2019).

[57] Greenblum and Kasperbauer (2018, p. 315).

with accounts offered by Savulescu[58] and Stahl and Emanuel[59]). The key feature of these accounts, emblematic of their classification in stage two of the discourse over conscience in healthcare, is that they conceive of professions as mere voluntary occupations. This is illustrated by Stahl and Emanuel's discussion of the difference between physicians and conscripts in war and by Greenblum and Kasperbauer's respect for the autonomy of the physician being satisfied by the physician simply choosing their occupation. This leads to the first point, which is that if and only if one believes physicians to be government (or semi-public) officials could Greenblum and Kasperbauer's argument be sound. If another description is more apt, then the argument fails.

A second concern has to do with the kind of work physicians do and with the practical judgment that they employ and exhibit. Arguing that "the full range of legitimate medical options" must be offered, like a number of the other points from Greenblum and Kasperbauer, is something that should and does garner much agreement, prima facie. Where more work is needed is in the articulation of those options and a realistic reflection on and of the daily practice of medicine. One assumes that the qualification of "legitimate" has been added to do this work, but there are—as physicians and other health professionals will share, if asked—a much greater variety of potential (and legitimate) options that are omitted in standard shared-decision-making conversations with patients due to medical professionals employing what they take to be their best judgment. The most common case in this area is the reduction of lists of potentially useful medications to one of a few because in the physicians' best judgment some might have less desirable side effects, the physician may have less experience with them, or simply because the list is quite long and the patient need not entertain all of them in order to receive potentially effective treatment options.

Third, as evidenced by earlier work in this book (see especially the second chapter), even though there are commonly held standards rooted in the best practices of medicine that are both a good and needed feature of the practices of any health profession, there does exist disagreement among professionals. It is not clear that government employees (not politicians, but the kinds of employees that Greenblum and Kasperbauer have in mind here) exist in a profession which does admit of disagreement. Positions on conscience should account for this fact. In some ways, this point also speaks to the rejection of the possibility of appeals to conscience due to reasons of predictability—the idea that consistency is needed in order for patients and others to have reasonable expectations when visiting one as opposed to another physician. Put to the side the flaw in this line of thinking that all physicians are equally skilled. What proponents of positions motivated by concerns about reasonable expectations hope for are the satisfaction of clear social interests in being able to predict outcomes, most especially in terms of efficiency. However, consider the aforementioned hypothetical case wherein the physician did not refer a patient for MAID due to considerations of their conscience. As illustrated by that case, if the proper processes are in place and followed, the physician would not be treating the

[58] Savulescu (2006).

[59] Stahl and Emanuel (2017).

patient as a social unequal; similarly, if such processes are in place, then inefficiencies in the system may be quite minimal.

These three considerations are jointly sufficient to raise concern about the accuracy of the positions in stage two: in giving up the possibility of comprise and moving away from attempts to resolve disagreements between persons by means of reflection on a singular (or at least two more closely related) conceptions of autonomy, positions in stage two of the dialectic differ—and more so than they may have first appeared—from those in stage one and from the daily practice of medicine. Now, let us consider the final objection against the representative example of stage two, which is decisive.

The aforementioned decisive objection comes from an exchange in which Greenblum and Kasperbauer attempt to respond to a critique of stage one positions, that they believe applies to their account, as well. Citing an earlier analysis[60] of Aulisio and Arora's view about the failure to take seriously practitioner moral judgment—though different interpretations of that critique exist—Greenblum and Kasperbauer argue the following:

> In response, we would point out that in a liberal democracy people disagree about many important moral questions, including considerations of moral complicity and responsibility. Both we and Aulisio and Aurora could respond to Pilkington by pointing out that the physician's understanding of moral complicity is simply part of his or her own value system. The physician is welcome to live in accord with those values, but they are forbidden from forcing patients to live by those values.

Placing interpretive disagreements to the side, this exchange is illustrative of the similarities and differences between positions at each stage of this dialectic. In seeking to respond to the critique about the rich and complex lives of persons—something which, even if Aulisio and Arora's view failed to do justice to, it at least took seriously—because persons are not a category in Greenblum and Kasperbauer's analysis, they need to reinterpret the critique in terms of physician's moral judgment, as opposed to a rejection of the view of persons as mere lists of discrete values. But, even in so doing and avoiding the more challenging objection, their response fails. To allow physicians to live according to their own value system—borrowing from the value imposition concerns of Aulisio and Aurora—but forbid them to force patients to live by those values is not what is at issue. Once they place considerations of moral complicity within their category of "values," the things according to which they are happy to let physicians live, an inconsistency arises because it is this very consideration (i.e., whether persons have the freedom to choose and adopt their own account of moral responsibility) that is at issue in the argument. This leads to three important conclusions: First, debates about conscience, at least those illustrated within the staged dialectic articulated in this text, truly are about moral responsibility. This gives further credence to arguments in Chaps. 2 and 3, especially those in Chap. 3 highlighting religion as a red herring in this debate. Second, better (at least more sensible and more clearly articulated) positions than Greenblum and Kasperbauer's are those that simply bite the bullet and reject conscientious practice *of any type* in medicine—likening physicians to assembly line workers optimizing

---

60 Pilkington (2019).

widgets (granted, very complicated widgets and very complex assembly lines). Third, it leads us to the third (and final stage) of this particular telling of the conscience dialectic: the inclusion of moral responsibility and the need for public philosophical efforts to work out admissible accounts of moral responsibility. It is to that discussion that we now turn.

4.3 Stage Three: The Challenge of Moral Responsibility in Practice and the Need for a Return of Public Philosophy

The suggestion of a public philosophical exchange as both the right response (and the right forum) to address differences in understandings of conscience in medicine may appear wrongheaded. The idea that further conversation about issues around which there are great disagreements risks simply repeating the problem and doing so in a way which—given the nature of certain kinds of philosophical discourse—may fail to heed the urgency of health situations that would be impacted by these considerations.

Two points are relevant in support of such a conversation and in response to these appropriate concerns. First, this book is a work in the field of bioethics, or at least one which engages bioethics. This field has, of late and with increasing frequency, recognized the importance of public engagement around bioethical issues.[61] Second, bioethics as a multidisciplinary endeavor has, since its inception engaged a variety of disciplines and perspectives. Recent work has suggested that bioethics professionals may be less diverse than the disciplinary origins of the field would suggest,[62] but at its core and at its best, it engages all those who are touched by ethical concerns related to health. That said, the analysis of issues related to conscience in this text does not fit standard approaches to issues in bioethics. Bioethics as a field in which values are adjudicated between and in light of deference to the four principles (with significant weight given to autonomy[63]) fittingly requires an explanation of the fit of the approach in this text: why recommend a broader public philosophical conversation as the needed venue to engage these ideas, and not simply a healthcare ethics committee meeting?[64] This is for a few reasons, which this section takes up. It does so by means

[61] Consider that the leading bioethics organization in the U.S. made this the focus of a recent annual meeting.

[62] See, for example, the results of a recent survey of bioethicists: Pierson et al. (2024).

[63] For a recent, practical work in bioethics that illustrates this, see: Post and Blustein (2015). It is worth noting that one might object to this description of bioethics as focused on autonomy, highlighting that the field is more diverse. First, see the preceding footnote. Second, even if it is true that preferences for autonomy in bioethics are not as strong as the used to be, it remains one of the most influential concepts in the work of CDs and so it is worth considering.

[64] For a review of (and critique of) one of the leading healthcare ethics committee handbooks, see: Pilkington (2018b).

of an analysis of contemporary bioethics through a lens provided by the work of philosopher Alasdair MacIntyre.

In the last few decades,[65] a good deal of ethical reflection on medicine, in particular, and healthcare, more generally, has come from persons who hold the professional title of "bioethicist." This might have struck leaders of ethical reflection on medical practice and healthcare at the dawn of bioethics,[66] like a Paul Ramsey or a Richard McCormick, as odd. It is no surprise that theologians, as well as philosophers, legal theorists, and social scientists are interested in professional practices and ethical standards arising from the intimate and complex issues associated with human health and the wellness of members of their communities. Understood in this way, the line of inquiry that bioethics appears to encompass—ethical questions about health and health-related practices—can be traced back much further, to Aquinas, to Aristotle, to Socrates, and many, many others who would be fitting forebearers of public philosophical discourse on the subject.[67] This is because the subject matter engages foundational questions of our shared human experience. What would have been surprising to a Ramsey or a McCormick is the recent rise and dominance of the *profession* of bioethics. Bioethicists, so understood, are neither philosophers nor lawyers applying the resources of their discipline to complex ethical issues in medicine, but rather they are independent members of the relevant ethical task spaces in healthcare—governing much of its ethical conversation. What once was a multidisciplinary endeavor is now, at least in some of the functions taken to be at the field's core, often becoming the task of a single profession.[68]

The rise of *this* sovereign profession (to borrow from a social scientist writing on medicine almost 40 years ago[69]), how it came about, and how it became the dominant discipline within medical and healthcare settings regarding questions of ethics is both interesting and important.[70] However, the aims of this section are more modest. To situate the relevant arguments in this section and explain why a public philosophical

[65] The argument in this section requires a brief historical framing. The aim of this section, as well as the chapter and the book, overall, is not to offer a history of the field of bioethics. As such, it does not engage different narratives or argue for a particular accounting of the history of the field as opposed to another. Rather, the historical framing aims to: (a) give to the general reader some context for the arguments in the chapter and (b) for readers who are more immersed in the field of bioethics, the hope is that this short historical framing will bring to mind the strengths and weaknesses that come with this particular historical accounting, again, adding context to the arguments.

[66] Like the representative examples and stages addressed earlier in this chapter, there are different histories of bioethics. This is one, but another could have focused on other contributors, such as Dan Brock, or Tom Beauchamp and James Childress.

[67] And this is only to mention the names of influential philosophers to underscore the point that one accounting of bioethics extends it much further than the common discussions that arose, arguably, 60 years ago.

[68] It is interesting to put this notion in conversation with the recent survey of bioethicists, the results of which suggest bioethicists are much more homogenous than is sometimes suggested (Pierson et al. 2024).

[69] Starr (2017).

[70] See Evans (2014). For a different perspective on the effective task spaces of bioethics, especially as they relate to research, see: Elliott (2024).

conversation is the right response, as opposed to a more common bioethical analysis, the text explores why a new analysis is needed through an assessment of bioethics that relies on MacIntyre's concept of a "character." Ultimately, it is argued that the bioethicist is a character of the MacIntyrean sense within modern medicine. To this end, this section proceeds with a sketch of the character of the bioethicist. Then with a description of the features of MacIntyre's characters that are helpful in thinking about the bioethicist, in particular: (1) moral constraint and the fusion of role and personality, (2) characters as moral representatives of their culture, and (3) external imposition on characters. After which, some objections from within the philosophical literature on MacIntyre's characters are considered; primarily the concern that "the bioethicist" is really just a version of the "manager" or the "therapist." The section concludes by addressing the implications of this analysis for conscience in medicine and highlights why the public philosophical approach is needed, as opposed to simply keeping the assessment in the halls of bioethics analysis.[71]

4.3.1 Bioethicist as a MacIntyrean Character

As noted above, many point to the late 1960s or early 1970s as the start of the discipline of bioethics. At this time, so the story goes (though surely inflected with a bit of nostalgia), physicians and other healthcare practitioners are interacting with practitioners of other disciplines (e.g., theologians, philosophers, and social scientists). There are figures like Ramsey, shadowing physicians at Georgetown to better understand medicine and setting up multidisciplinary conferences to bring together academics of diverse disciplines and members of the healthcare professions to discuss the challenging medical ethical issues of the day. Today, though the bioethics terrain is still a bit murky, what was once a multi-disciplinary endeavor has become, to a large degree, a singular field's arena.

For the purposes of this section, bioethics is to be understood broadly as a field that addresses ethical issues about health and practices related to health and medicine. In the view expressed in this chapter, bioethics is understood as a necessarily collaborative endeavor; that is to say that one cannot do it or, at least, one cannot do it well by oneself. It requires participation from members of a variety of professions (or disciplines, fields, subject matters, et cetera) because it addresses issues proper to the domains of those professions (and disciplines, fields, subject matters, et cetera).

This can be illustrated well through considerations of some of the kinds of questions that bioethicists take up. They consider questions, including those about the ethical implications of emerging medical technologies, such as: Should we turn off a

[71] Before beginning the argument, it is important to note that this is a particular kind of analysis and one that may be less familiar to some working within bioethics, given their particular disciplinary home. Critiques of this approach, such as it lacks empirical proof or that it offers merely a caricature of a bioethicist are, at a higher level of abstraction, well-taken, but miss the point of this line of inquiry. Such critiques belong in discussions about methodology in multidisciplinary fields with respect to the home disciplines of such a field's diverse interlocutors.

pacemaker in an elderly patient because of the request of a family member, knowing that the patient is completely dependent on that device? Should we distinguish pacemakers and implantable cardioverter defibrillators when so deactivating? Should artificial intelligence be relied on for clinical decision support? How ought equity be ensured in the rollout out of AI platforms when knowledge of the decision-impacting inputs is obscured? Bioethicists also ask questions about the practice of medicine, such as: What is owed to a patient: honesty? the best course of treatment? the greatest possible exercise of autonomy? What should be done if these conflict? Bioethicists ask questions about justice within health care systems, both locally and globally, with reflection of different communities and societies within broader collections of persons and conceptions of health; that is, they ask general questions, like, "How should we allocate scarce medical resources: to those who can pay? To all citizens? To all persons living within our communities?" as well as particular practice questions, such as, "Is it morally acceptable to restrain, that is, handcuff, incarcerated women during labor and childbirth?".

A bioethicist could not weigh the issues at the heart of these questions and the ethical features of the questions themselves, without knowledge of what is going on, and in bioethics the answer to "what is going on?" in most cases is informed by social and political features, as well as features rooted in medical science. Put another way, issues that arise within practices associated with health often touch on a variety of areas of expertise; hence, the reliance of the arguments in this chapter on the importance of different perspectives in "doing" bioethics, as well as the resulting conclusion that a public philosophical conversation is needed to address considerations of conscience in medicine.

If this account of bioethics is correct or, at least, if one takes it seriously as a plausible interpretation, one will see yet another reason why a public philosophical conversation is the right resolution: because bioethics engages normative questions, both theoretical and applied, associated with the health of human beings, it is not surprising that bioethics is often thought of as a proper subdiscipline of philosophy.[72] Such an understanding of bioethics, however, suggests an alternative history, one which does not coalesce in the 1960s or 1970s, but rather is part of an intertwining set of conversations, including those about health, medicine, what it is to be human, and ethics. Attending to that history, as noted above, one could trace questions of health and ethics back to Aquinas or Aristotle, to Socrates, or to many others.[73]

[72] This a contested claim and that is for at least two reasons: First, some would place it elsewhere, theology or law; second, given the rise of bioethics MA and PhD programs, some would argue that bioethics is distinct from any other discipline.

[73] The content of bioethics as it is currently practiced, is deeply influenced and, normatively shaped by *the* four principles; though in recent years work in bioethics has articulated other viewpoints and some of those have gained traction, critiquing this kind of principlist approach in favor of virtue-based approaches or, simply in favor of the employment of other concepts, such as human dignity. Thus, there are arguments about the content of bioethics and disagreement about emphasis, in addition to questions of individuation and who has proper standing to say what, and where—the landscape is murky.

Reflection on this history and narrative of bioethics, raises questions about this important space for ethical reflection on health and healthcare: Who should be in this conversation? Who has standing? In light of the claim that bioethics is necessarily collaborative, consider two competing approaches—call them Person A and Person B.

Person A engages bioethical issues by applying an interdisciplinary approach. (Though this is a caricature to some degree, it is fair to attribute this approach to many professional bioethicists working within the field today). The approach of Person A has a few key features, foremost of which is that it does not rely on another discipline as primary—bioethics is *the* discipline. Or, you might say that this approach requires engagement with diverse disciplines in such a pervasive and foundational manner as to have led to the creation of a new one.[74] Consider, for example, that professional bioethicists today might hold degrees, not in law or philosophy or a social science, but rather possess MAs, PhDs, or other advanced certificates *in* bioethics. One might also note the recent push and now successful program within the United States of credentialing bioethicists who perform a particular set of clinical tasks, described under the umbrella of clinical ethics consultation. The HEC-C designation, supported by the foremost bioethics organization in the U.S. appears, at least anecdotally, to be increasingly widespread, though many clinical ethicists still practice without such a credential.

Contrast Person A with another, Person B. Person B employs a multidisciplinary approach. This is an approach that, instead of conceiving of bioethics as the practitioner's primary disciplinary home, relies for its foundations on another primary discipline, using the tools and bodies of knowledge of those ways of approaching the world, to reflect on, engage, or—most significantly—to "do" bioethics. To clarify Person B's approach, think of persons who are originally trained in, for example, the discipline of philosophy and for whom the field of bioethics is an interesting and important subfield (properly described as applied ethics, a subset of ethics) of the discipline of philosophy. As such, Person B might hold a PhD or a DPhil in moral philosophy. They might also hold degrees in theology or feminist political thought (or both); Person B is, in many ways, a much more diverse amalgamation of training histories. The approach of Person B requires engagement with the resources of a variety of disciplines but retains a primary orientation or dominant discipline. These trainings make their way into bioethical discourse, which is to say that they matter and have downstream effects. For example, if one understands bioethics to be a subset of philosophy, in light of one's training, then the "rules" of philosophy apply to bioethics. This would mean that if one were to do bioethics, one must offer an argument. In fact, it might even be understood that the offering of arguments and the critique of them as ways to move thought (and the conversation) forward are marks of respect among Persons B. A similar description and application of training could

[74] For a more extensive treatment of the ideas of multidisciplinary, interdisciplinary, and related concepts, as they might apply to bioethics, see Pilkington (2018a). Some of the description in this subsection mirrors thoughts expressed there.

be offered for political theorist doing bioethics, or lawyers, or theologians, or many others. The key to counting as Person B is that one must respect multidisciplinarity.

Person B ought to be someone well-versed in a particular practice and tradition of the study of aspects of our humanity who is then able to look at a distinct practice and bring resources of analysis to that practice that might not have been utilized by the latter practice itself. Person A, in contrast, might very well be a lone bioethicist in a hospital setting, doing clinical ethics consultations relying on practical claims rooted in the four principles. Person A embodies an interdisciplinary approach in that the relevant disciplinary expertise is embodied in that individual, *the bioethicist*. Even though this dominant approach to bioethics focuses on clinical ethics, it has filtered into, and informs discussions of, policy and debates within academic bioethics with an impact well beyond the clinic.

4.3.2 Characters

With the above sketch of a bioethicist in mind, let us turn to the conception of a "character." The features of MacIntyre's characters that are essential for this analysis are three: (1) moral constraint and role/personality fusion, (2) moral representation of culture, and (3) external imposition on characters. I take up each, in turn.

First, characters are "… a very special type of social role which places a certain kind of moral constraint on the personality of those who inhabit them in a way in which many other social roles do not."[75] Further, MacIntyre adds, "In the case of a *character* role and personality fuse in a more specific way than in general; in the case of a *character* the possibilities of action are defined in a more limited way than in general."[76] The contemporary bioethicist is constrained in their thinking, being governed not merely by the four principles, but a particular and fairly simplistic account of one of these principles as their true guide. As Linda Farber Post and Jeffrey Blustein have recently written,

> Autonomy is the ethical principle widely considered most central to health care decision making. Its prominence here and in other bioethics literature reflects the heightened emphasis typically accorded patient rights, self-governance, and individual choice…Only when the individual cannot make decisions are others asked to choose. Autonomy gives priority to personal values and wishes…[77]

This focus on autonomy as, essentially, patient preference has the benefit of being action-guiding: by and large, do what the patient requests. The bioethicist reinforces this narrative, but understanding what it is that is requested—especially in light of complex procedures, varying degrees of health-related knowledge, and the uncertainty of outcomes—is not always clear. This leaves the bioethicist in a position to facilitate, translate, or support a dialectic between the "relevant stakeholders."

---

[75] MacIntyre (2007, p. 27).

[76] MacIntyre (2007, p. 28).

[77] Post and Blustein (2015).

The translation is possible because a unit of measure exists: values. Recall the aforementioned considerations of values and debates over conscience rooted in autonomy. Many CDs would fit in the category of Person As. Though, again, a bit of a caricature, on this account, persons have—or *are*—lists of values. Again, recall the arguments from earlier in this chapter. The bioethicist helps to navigate these sets of values—not by supporting significant self-interrogation or the usual ways one might think about practically reasoning together with others, but by placing these sets of values in contact with other sets of values—those of the physician, those of the nurse, the hospital, the health system, et cetera. In fact, this is exactly what some analyses of conscience in medicine exemplify. Consider Greenblum and Kasperbaur's framing of conscience as a values conflict in which the patient always wins out. Understanding persons in this way might be helpful for the necessary bioethics tasks, because in addition to simplifying the interactions in a complex space, they also allow for the adjudication of disagreement.[78] It is unsurprising that disagreement occurs within the healthcare space, as health is something which is deeply prized by many, if not all, of us, and something about which we possess different views, understandings, and meaning-making accounts. Matters which affect our health affect our way of life and how we respond to life-affecting situations impacts our self-conceptions, and deeply. The bioethicist boils this down to a set of values and facilitates comparisons of such sets when conflict arises.[79]

A second important feature of characters, as MacIntyre construes them, is that they are moral representatives[80] of their culture: "They...are so because of the way in which moral and metaphysical ideas and theories assume through them an embodied existence in the social world. *Characters* are the masks worn by moral philosophies."[81] The bioethicist represents the current medical and healthcare culture. They embody an ethic that allows in a variety of moral points of view, many of which are in conflict. She might, for example, feel a tension between experiences in three clinical ethics consultations. First, a situation in which a physician, begrudgingly, meets with a family who wishes for further care which he is refusing as the patient "didn't have

[78] Though this approach fails, at least as a way to understand conscience in medicine, this does not detract from the important and laudable work that is done in clinical ethics around garnering agreement with respect to patients, families, other loved ones, care teams, and hospital systems. The conflict resolution that occurs in these spaces in some very, very challenging contexts involves amazing work. However, the practices of clinical ethics consultation, often a version of conflict resolution and sometimes even of restorative justice, is not the same thing as policy creation or practices which challenge and restrict the conscientious practice of individual physicians.

[79] It should be noted that this character is, in many ways a caricature, but one which carries truth with it. It does not, of course, represent all bioethicists—that is not what a MacIntyrean character would do or be in the first place. It is also not a characterization of Person B, but of Person A.

[80] MacIntyre writes, "One of the key differences between cultures is in the extent to which roles are *characters,* but what is specific to each culture is in large and central part what is specific to its stock of *characters.* So the culture of Victorian England was partially defined by the *characters* of the Public School Headmaster, the Explorer and the Engineer; and that of Wilhelmine Germany was similarly defined by such *characters* as those of the Prussian Officer, the Professor and the Social Democrat" (MacIntyre 2007).

[81] MacIntyre (2007, p. 28).

a shot." Second, a situation in which a patient requests medication to end her life but the doctor "didn't' get into medicine to kill people." Finally, a third situation in which the hospital policy specifies that because a patient had died, care should be removed, but the family of the patient views him as still living and the organ procurement team is politely knocking. Navigating these conflicts and adjudicating these disagreements, many of which involve value-laden expressions offered by well-meaning but beleaguered clinicians, family members, and patients is no easy task. That these experiences did not yield the resources to adjudicate disagreement around the topic of conscience in medicine should not be surprising. This may be why, for instance, the common approach encapsulated by Greenblum and Kasperbaur's account opts for the simple resolution of disagreement by ridding the conversation of one of the parties. The bioethicist only has the moral and linguistic tools to embrace simple preference-satisfaction because they exist in a world in which there are answers (e.g., the "science of medicine" informing hospital policy specifies when a patient is dead, regardless of whether their heart is beating and their lungs are moving up and down, even if aided by machines) and in which there are members of an amorphous set known as "values" which must stack up against the hard evidence. Adjudicating such conflicts without appeal to broader resources is challenging. The bioethicist's unit of measure—*values*—and thinking about persons as lists of values illustrates their satisfaction of this feature of a *character*. They represent an embodiment of neutrality, adjudicating between different lists of "values plus medical facts," also known as patients, and "values plus professional facts," also known as physicians.[82]

"A *character* is an object of regard by the members of the culture generally or by some significant segment of them. He furnishes them with a cultural and moral ideal,"[83] argues MacIntyre. The bioethicist represents *good* that is done by health professions and institutions; they are appealed to for a "stamp of approval" or in an earnest way as an ethical expert on healthcare matters. They may be an expert, but an expert in light of the neutral, "values"-based system described above. Success is resolution of tensions among healthcare stakeholders, which keeps the industry of healthcare moving forward—both with goods internal to the practice of medicine and to health being realized, but also with obstacles being places in the paths of achieving them due to institutions aiming at external goods. The healthcare enterprise can claim to maintain high moral standards and its members can hold themselves in high ethical regard because what is being done is certified by the bioethicist.[84]

We have already seen inklings of the third feature of characters, which the bioethicist satisfies, they are externally imposed upon. MacIntyre writes:

---

[82] Note that a bioethicist—not a MacIntyrean character—but a bioethicist working in ideas from stage one, would have further resources, thinking of physicians and patients as persons.

[83] MacIntyre (2007, p. 29).

[84] Enabling the health enterprise to move forward is in many, many ways both good and needed. The worry for the bioethicist—the practitioner, not the character—is whether this work keeps the hard questions, the deep questions, the existential questions that violations of health raise from being placed at the fore? Are they relegated thanks to the work of bioethicist and their framework of neutral adjudication of lists of values or their roles in institutions? For examples of this concern realized, see Elliott (2024), and attend to the examples where whistleblowers mention bioethicists.

> There are then many cases where there is a certain distance between role and individual and where consequently a variety of degrees of doubt, compromise, interpretation or cynicism may mediate the relationship of individual to role. With what I have called *characters* it is quite otherwise; and the difference arises from the fact that the requirements of a *character* are imposed from the outside, from the way in which others regard and use *characters* to understand and to evaluate themselves.[85]

Healthcare professionals appeal to bioethicists, institutionally and individually, as the bar for ethical practice. This is not to say all medical professionals merely seek a stamp of approval, some might define themselves in opposition to these institutionally supported moral arbiters. Yet, bioethicists inform hospital policy, serve as watchdogs,[86] and can be consulted on any case to assist in adjudication of disagreements and ensure the ethical quality of a decision (albeit constrained by the aforementioned resources). Bioethicists are also embedded in powerful institutions, both in academic medicine and in other hospitals and health systems), must deal with social and political issues that make their way into medicine, and address the daily challenges (for example, the moral distress that acutely arises in a pandemic) of those working in healthcare. They are reactive creations.

4.3.3 Rejection of the Analysis: Too Many Characters, All the Same Character

There are at least three objections to the arguments in the preceding subsections. First, one might reject my interpretation of MacIntyre's character and its use in analyzing the arena in which much of the debate over conscience in medicine occurs: bioethics. That objection might claim that I have borrowed the manager, put a whitecoat on them, and armed them with some medically altered version of bureaucratic rationality and a failure to evaluate ends in medicine. A second objection is that the discussion of the characters and healthcare is repetitive, as it, at best, shows that (merely) a new way to talk about the *therapist* was found. A third objection to consider is that the character described is too much of a caricature and that there exist philosophers, lawyers, and others working in bioethics, as well as moves within that field (e.g., narrative approaches) which suggest a greater degree of multidisciplinary work and deeper resources that have been ignored by the analysis.

---

[85] MacIntyre (2007, p. 29).

[86] Some argue they perform this function well (See, for example, Evans [2014]; others argue that they have failed, see, for example, Elliott [2024]).

4.3.4 Some Replies: Characters of Time, Place, and "Practice"

The objections of the previous section are well taken. The first two objections are likely to be offered by careful readers of MacIntyre; the final objection is likely to be offered by someone working in bioethics. Let us take each objection in turn. First, similarities between the characters—the bioethicist and MacIntyre's three—surely exist. This is not because the character analysis is repetitive but because 1981 is not that far from today, and the advent of bioethics the decade or so before the publication of *After Virtue* set the conceptual resources and framed the creation of the field in a particular direction. Though similarities surely exist, the bioethicist is not merely a healthcare manager. Midlevel administration members in hospitals might fit that description, applying bureaucratic rationality to the field of healthcare, but the bioethicist is distinct. The bioethicist has embraced neutrality, and embraced it in a deep way, but that neutrality is coupled with a particular account of autonomy in medical decision-making, and a metaphysics in which persons are lists of values (plus some medical facts). The bioethicist does not lack ends or block the critical evaluation of ends; the conceptual resources that they have do not allow for deeper engagement with important questions relevant to health. They are not merely pushing away those ends; the questions are answered by their limited framework.

Second, contrary to the concern raised in the second objection, the bioethicist and the therapist are distinct. The bioethicist would, like the therapist, treat the ends of patients as given. However, the distinctive feature of the bioethicist is their view of the medical-ethical world as lists of values clashing, and through facilitation within a framework of autonomy, medicine can be practiced ethically in spite of disagreement. The bioethicist does engage in moral debate, it is just significantly limited. That they are open to such debate illustrates a difference between them and the therapist. It is true that the bioethicist, as a character constructed within a broadly emotivist culture, might also embody the obliteration of the distinction between manipulative and non-manipulative social relations. That is, implicit in the bioethicist's understanding of persons as lists of values and the manner in which they seek to adjudicate disagreement, a stark line between manipulating and not manipulating is no longer drawn. However, noticing this similarity, like the others, is not sufficient to reject my claim. Similarities are to be expected.

Reflecting on the origin of the character, it may be that the bioethicist is what comes about from the mixing of an emotivist culture with the "hard facts" of medical science. The bioethicist avoids incommensurability (even if wrongly) with a focus on autonomy and a conception of human beings as lists of values. There is an impersonal nature to her arguments, in light of this—and in fitting with MacIntyre's own characters—but those arguments do often include personal content (e.g., "Mrs. Jones values x, so…") and the encouragement of personal arguments (e.g., "Explain your values, Mr. Smith, and please fill out an advanced directive").

In response to the third objection, the one likely posed by someone in or very familiar with the field of bioethics, my character surely is a caricature, but a kernel of truth exists even in caricatures. The field of bioethics would do well to reflect on this character, as well as those in administrative and other positions within hospitals and healthcare systems. If pushes toward narrative medicine continue, broader and more inclusive conversations about human beings and human goods are embraced, and governing principles are expanded, the aforementioned concerns about the way that some professionals practice bioethics would dissipate. If this becomes the case, then we might end up in a place where physicians and nurses, techs and administrators, philosophers and theologians come together to discuss challenging situations in healthcare and medicine, and that is where my hope lies. If that conversation is had in public with the understanding that careful argumentation, clear and concise analysis, and aims of consistency will help us get it right, then a public philosophical discussion might be possible as a way to address issues of conscience in medicine.

4.4 Conclusion

In the final section of the previous chapter and in this chapter, I have argued that conceptions of moral responsibility are fitting subject matter for respect for conscientious practice in medicine. This shifts the focus of conversations in this space, away from more common bioethics analyses which pit professional autonomy against patient autonomy, understand appeals to conscience as only protected religious-based exceptions to reasonable expectations, or must fit into wartime accounts of CO (properly understood). To do this, I described the kind of moral responsibility under discussion, replied to autonomy-focused arguments in this space, and affirmed (and reaffirmed) analyses of the concepts of conscientious practice. Recognizing that the common arena for this broader conversation is the field of bioethics—in particular clinical ethics consultation and healthcare ethics committee spaces—I described why the resources of contemporary bioethics may not make it the right place for such a conversation through reflection on the work of a prominent ethicist, Alasdair MacIntyre, ultimately recommending a public philosophical conversation.

If the arguments in this chapter are correct, then respect for the diversity of accounts of moral responsibility as part of the rich lives and self-conceptions of persons is needed and, at the very least, is admissible. In the next chapter, I turn to the actions performed in medicine, engaging the role of the medical professionals, as well as an importantly new influence on medical actions—which is proving impactful at this distinct and pivotal juncture—the role of artificial intelligence in medicine.

References

Aulisio, Mark P., and Kavita Shah Arora. 2014. "Speak No Evil? Conscience and the Duty to Inform, Refer or Transfer Care." *HEC Forum: An Interdisciplinary Journal on Hospitals' Ethical and Legal Issues* 26 (3) (September): 257–266. https://doi.org/10.1007/s10730-014-9242-8.

Baldwin, James. 1993. *The Fire Next Time.* 1st Vintage International ed. New York: Vintage International.

Beauchamp, T. L., and J. F. Childress. 2013. *Principles of Biomedical Ethics.* 7th ed. Oxford University Press.

Brandt, A. 1978. Racism and Research: The Case of the Tuskegee Syphilis Study, *The Hastings Center Report*, Vol. 8, No. 6.

Cavanaugh, T. 2018. *Hippocrates' Oath and Asclepius' Snake: The Birth of the Medical Profession.* Oxford University Press.

Cummins, Paul J. 2021. "Conscientious Objection and Physician–Employees." *HEC Forum* 33 (3) (September): 247–268. https://doi.org/10.1007/s10730-019-09390-8.

Department of Health, Education, and Welfare and National Commission for the Protection of Human Subjects of Biomedical and Behavioral Research. 2014. "The Belmont Report. Ethical Principles and Guidelines for the Protection of Human Subjects of Research." *The Journal of the American College of Dentists* 81 (3): 4–13.

Doernberg, Samuel, and Robert Truog. 2023. "Spheres of Morality: The Ethical Codes of the Medical Profession." *The American Journal of Bioethics* 23 (12) (December 2): 8–22. https://doi.org/10.1080/15265161.2022.2160514.

Elliott, Carl. 2024. *The Occasional Human Sacrifice: Medical Experimentation and the Price of Saying No.* 1st ed. New York, NY: W.W. Norton & Company.

Engelhardt, H. T. 1989. "Freedom vs. Best Interest: A Conflict at the Roots of Health Care." In Dax's Case: Essays in Medical Ethics and Human Meaning, edited by. L. Kliever, 79–96. Dallas, TX: Southern Methodist University Press.

Evans, John Hyde. 2014. *The History and Future of Bioethics: A Sociological View.* 1. issued as paperback. Oxford: Oxford University Press.

Farber Post, L. and J. Blustein. 2015. *Handbook for Health Care Ethics Committees.* JHU Press.

Flannery, Kevin L. 2001. *Acts Amid Precepts: The Aristotelian Logical Structure of Thomas Aquinas's Moral Theory.* Washington, DC: Catholic University of America Press.

Flannery, Kevin L. 2012. *Christian and Moral Action.* The Institute for the Psychological Sciences Monograph Series, Vol. 3. Arlington, Virginia: The Institute for the Psychological Sciences Press.

Greenblum, Jake, and T. J. Kasperbauer. 2018. "Forget Evil: Autonomy, the Physician–Patient Relationship, and the Duty to Refer." *Journal of Bioethical Inquiry* 15 (3) (September): 313–317. https://doi.org/10.1007/s11673-018-9854-9.

Johnson. K. 2019. *Medical Stigmata: Race, Medicine, and the Pursuit of Theological Liberation,*

Kaldjian, Lauris C., and Bryan C. Pilkington. 2021. "Why Truthfulness Is the First of the Virtues." *The American Journal of Bioethics* 21 (5) (May 4): 36–38. https://doi.org/10.1080/15265161.2021.1906991.

Kant, I. 1785. *Groundwork for the Metaphysics of Morals.*

Kilbride, M., and S. Joffe. 2018. The New Age of Patient Autonomy: Implications for the Patient-Physician Relationship. *JAMA* 320 (19): 1973–1974.

MacIntyre, Alasdair C. 2007. *After Virtue: A Study in Moral Theory.* 3rd ed. Notre Dame: University of Notre Dame Press.

MacIntyre, Alasdair. 1984. *Is Patriotism a Virtue?* (The Lindley Lecture). Lawrence: University of Kansas. Reprinted in Primoratz (ed.), 2002.

MacIntyre, Alasdair. 1988. *Whose Justice? Which Rationality?* Notre Dame, IN: University of Notre Dame Press.

May, T. 2002. *Bioethics in a Liberal Society: The Political Framework for Bioethics Decision Making.* Baltimore: Johns Hopkins University Press.

May, William F. 2000. *The Physician's Covenant: Images of the Healer in Medical Ethics.* 2nd ed. Lexington, KY: Westminster John Knox Press.

Mellema, Gregory. 2016. *Complicity and Moral Accountability.* Notre Dame: University of Notre Dame Press.

Meyers, Christopher. 2021. "Deception and the Clinical Ethicist." *The American Journal of Bioethics* 21 (5) (May): 4–12. https://doi.org/10.1080/15265161.2020.1863513.

National Commission for the Protection of Human Subjects of Biomedical and Behavioral Research. 1979. *The Belmont Report: Ethical Principles and Guidelines for the Protection of Human Subjects of Research.* Retrieved from https://www.hhs.gov/ohrp/regulations-and-policy/belmont-report/read-the-belmont-report/index.html.

Pierson, Leah, Sophie Gibert, Leila Orszag, Haley K. Sullivan, Rachel Yuexin Fei, Govind Persad, and Emily A. Largent. 2024. "Bioethicists Today: Results of the Views in Bioethics Survey." *The American Journal of Bioethics* (May 6): 1–16. https://doi.org/10.1080/15265161.2024.2337425.

Pilkington, B. C. 2014. "On Omissions and Artificial Hydration and Nutrition." *Journal of Medicine and Philosophy* 39 (4) (August 1): 430–443. https://doi.org/10.1093/jmp/jhu022.

Pilkington, Bryan C. 2016a. "Dignity, Health, and Membership: Who Counts as One of Us?" *The Journal of Medicine and Philosophy: A Forum for Bioethics and Philosophy of Medicine* 41 (2) (April 1): 115–129. https://doi.org/10.1093/jmp/jhw001.

Pilkington, Bryan C. 2016b. "Do No Evil: Unnoticed Assumptions in Accounts of Conscience Protection." *HEC Forum* 28 (1) (March): 1–10. https://doi.org/10.1007/s10730-015-9274-8.

Pilkington, Bryan C. 2018a. "The Fiftieth Anniversary of Patient as Person: Paul Ramsey's Groundbreaking Approach to Christian Bioethics." *Christian Bioethics: Non-Ecumenical Studies in Medical Morality* 24 (2) (June 26): 111–125. https://doi.org/10.1093/cb/cby003.

Pilkington, Bryan C. 2018b. "Are Handbooks Still Useful? Yes and No; It Depends on How You Use Them: Handbook for Health Care Ethics Committees, 2nd Edition. By Linda Farber Post and Jeffrey Blustein. Baltimore: Johns Hopkins University Press, 2015. Pp. Xiii + 413. $49.95." *HEC Forum* 30 (2) (June): 153–156. https://doi.org/10.1007/s10730-018-9347-6.

Pilkington, Bryan C. 2019. "Remember Evil: Remaining Assumptions In Autonomy-Based Accounts Of Conscience Protection." *Journal of Bioethical Inquiry* 16 (4) (December): 483–488. https://doi.org/10.1007/s11673-019-09949-7.

bPost, Linda Farber, and Jeffrey Blustein. 2015. *Handbook for Health Care Ethics Committees.* 2nd ed. Baltimore: Johns Hopkins University Press.

Ramsey, P. 1970. *Patient as Person: Explorations in Medical Ethics.* New Haven, CT: Yale University Press.

Savulescu, Julian. 2006. "Conscientious Objection in Medicine." *BMJ* 332 (7536) (February 4): 294–297. https://doi.org/10.1136/bmj.332.7536.294.

Smith, P. 2019. "Moral Status and the Care of Impaired Newborns: An African American Protestant Perspective."

Stahl, Ronit Y., and Ezekiel J. Emanuel. 2017. "Physicians, Not Conscripts—Conscientious Objection in Health Care." *New England Journal of Medicine* 376 (14) (2017): 1380–1385. https://doi.org/10.1056/NEJMsb1612472.

Starr, Paul. 2017. *The Social Transformation of American Medicine: The Rise of a Sovereign Profession & the Making of a Vast Industry: With a New Epilogue by the Author.* Updated edition. New York: Basic Books.

Taylor, Charles. 1979. "Atomism." In *Powers, Possessions and Freedom*, edited by Alkis Kontos, 39–62. University of Toronto Press. https://doi.org/10.3138/9781487589417-005.

Waldron, Jeremy. 2008. The Dignity of Groups. Acta Juridica, NYU School of Law, Public Law Research Paper No. 08-53, October 20, Available at SSRN: https://ssrn.com/abstract=1287174.

Waldron, Jeremy. 2012. *The Harm in Hate Speech.* The Oliver Wendell Holmes Lectures, 2009. Cambridge, MA: Harvard University Press.

Waldron, Jeremy. 2015. *Dignity, Rank, and Rights.* Edited by Meir Dan-Cohen. First issued as an Oxford University Press paperback. Berkeley Tanner Lectures. New York, NY: Oxford University Press.

Chapter 5
The Medical Act

This final chapter takes up the idea of the medical act and its connection to appeals to conscience. In so doing it is important to remind the reader that this book is a work in bioethics, and not in metaphysics or medical sociology. Within the wide variety of conversations both interdisciplinary and multidisciplinary given a home in the field of bioethics, the arguments of this book are best understood as instances of applied ethics, the ethics of medicine. Though "medical ethics" as a phrase has gone out of fashion, it remains the best descriptor of this text's content. Describing this text as a work in medical ethics, within bioethics, helps to highlight that it takes up an issue that is part of and important to medicine and the practice there of, and examines it from a critical point of view. That point of view is, if not philosophical, informed by philosophy. The book contains arguments, those arguments are constructed so as to not be inconsistent with each other, but also with an understanding of the messiness of a human practice like medicine. As part of a work in medical ethics, this chapter aims to offer a description of what medical professionals do when they make a judgment. In doing so, it situates those judgments in a number of contexts—with respect to patients, institutions, communities, and societies, as well as in relation to other systems (in particular, that of AI). The chapter closes with a reflection on moral disagreement and how we might, if the arguments in this book are sound, move the conversation forward.

Before turning to these discussions, it may be useful to track the argument of the book thus far. In the preceding chapter, moral responsibility and what things would look like if it were taken seriously in medicine were both discussed in relation to conscience. A variety of arguments were engaged, with a particular focus on autonomy-based accounts. The results of this work illustrated a need for a return to public philosophy. Much of the argumentative heavy lifting was completed through reliance on the work of Alasdair MacIntyre as a means to critically engage conscience in medicine, as well as the field in which many interlocutors who work on the topic reside, bioethics. In the prior chapter, I argued that religion should not be the focal point of debates about conscientious practice or CO. These arguments

© The Author(s), under exclusive license to Springer Nature Switzerland AG 2025

B. Pilkington, *The Medical Act: Conscientious Practice in a World of Dissention and Disagreement*, The International Library of Bioethics 114,
https://doi.org/10.1007/978-3-032-00048-4_5

about the role of religion in debates over conscience are necessary, arising out of a discussion, in Chap. 2, of disagreement. That discussion centered on conceptions of professions and their role in disagreement and in its resolution. Part of these latter arguments involved clarifying the philosophical approaches that conscientious dissenters relied on, but which misunderstand the theoretical resources to which they appeal. With these arguments in place, questions were raised about relevant arguments in the literature, as well as the unargued for assumption that the burden of proof is always on the practitioner who invokes conscience. Prior to this, the first chapter of the book laid out the distinction between CPs and CDs. In light further discussion throughout the intervening chapters, it is important to highlight that CPs find practitioner judgments—including their practical reasoning about the situations their patients bring to them and their own self-reflections on their place in caring for the health of those patients—to be important to the practice of medicine. In contrast, CDs do not, which makes sense given that they prioritize that which is external to the practitioner, the health professions themselves, as well as laws, or other cultural, social, financial and regulatory standards. With this brief summary of the overall argument of the text thus far in mind, let us now turn to the medical act.

5.1 What Medical Professionals Do When Making Judgments

By a medical act, I simply mean whatever it is that medical professionals do in their capacities as physicians. What follows is a rough and loose articulation, not a clear set of necessary and sufficient conditions. I anticipate there will be actions or situations in which a physician is involved wherein the best action description pushes back on this formulation, but this description is intended as a starting point, not the final word on the matter.

What is captured in a description of the medical act as "whatever" medical professionals do may seem, at first, to be too wide a catchall to be accurate or, if it is accurate, then it is so in a meaningless way. However, the expansiveness of the first part of the description is bounded by the latter part. A few clarifying points will prove useful. First, it is true that according to this description, medical acts are very numerous. When a surgeon makes a cut or a stitch in the service of improving a patient's health or easing their suffering and aiding palliation, that is a medical act. When a family physician, who is an expert in internal medicine, prescribes blood thinners for a patient after that surgery, given concerns about stroke, that is a medical act; and when that same physician orders blood work to be done, that request for additional information is also a medical act. These are clear instances of what a medical act is, and I doubt there would be much objection, but a medical act also includes the smaller and potentially unnoticed parts of a physician's routine that are part of their care for patients. Consider the fairly mundane and perhaps ethically uninteresting act of washing one's hands before seeing a patient, meeting with others on the care team, or as part of "scrubbing in" for surgery. These too are medical acts, as they are part of what a medical professional does in their capacity as a physician. This would not apply to all instances of handwashing, as it would to all instances of "scrubbing in." While on vacation from work, if a physician washes their hands in the course of their interactions with others or as they move about the day, this would not constitute a medical act. It is true that there are potential consequences were they to fail to do so—they might become ill and not be able to return to work to care for patients, but those kinds of cases do not fall under this description of medical acts. The set that includes those behaviors as medical is too wide. However, this does not mean that "scrubbing in" is the only relevant handwashing. Consider the germ theory of disease, and how relevant and important handwashing is for good care of patients.[1] The reason for focusing on handwashing is that it is, in some contexts, a medical act and in others, it is not.

It should not be drawn from the aforementioned examples that anything a physician does outside of the hospital, not during their "9-5" shift, or at a distance from the trappings of their medical work cannot be relevant. Keeping up one's skills is medically relevant, and physicians are responsible for this, but spending too much time away from their work and losing those skills is not a medical act.

It should also be noted that in describing the relevant phenomena as medical "acts," failures to act are not excluded—both praiseworthy and blameworthy ones. Omitting to offer certain kinds of treatment because they are understood to be futile or not beneficial to the patient in light of the patient's goals, values, or simply the basics of medical science, falls into the category, as well, and so medical act should be understood to include relevant omissions.

It should, finally, be noted that there is a category of actions which a physician might undertake as a physician but not in their capacity as a physician. Consider, for example, lobbying efforts for better compensation, more fair working environments, or work hour limits and the like. These are cases where physicians are acting as physicians, it is through their membership in the profession and their advocacy for their professional members that they do this work, but they are not acting in their capacity as a medical professional. Given this distinction, the critique that this account is prejudiced in favor of patient care as a defining factor of a physicians' capacity might be leveled.[2] It is a fair criticism of the position.[3]

This is a rough description of what constitutes a medical act. It is not a full description and it does not offer the necessary and sufficient conditions of a medical action. This is for a few reasons. First, the parts of the medical act that are considered are those that bear on the focus of this text: conscience. Second, there is, admittedly, a vagueness in this description and bright lines are not drawn between some actions

---

[1] For an excellent illustration of the importance of handwashing, see CDC (2024).

[2] For an excellent discussion of the different roles that physicians might occupy and the norms that apply in each sphere, see Doernberg and Truog (2023). Doernberg and Truog do not prejudice the clinical space in their accounting of physician's roles.

[3] For a response to Doernberg and Truog's argument that focuses on the primacy of the clinical space and gives further background which informs my position, see Pilkington (2023).

and others. Lobbying for better working conditions as a physician is surely something a physician does, but it is not a medical act. How should we understand lobbying for resources for better health literacy or for vaccinations to make their way more quickly to a particular community when it is done by a physician? These actions are aimed at better health for patients. Yes, so too might better working hours, if those restrictions lead to the reduction of medical errors and, in turn, better patient outcomes. Lines must be drawn, even if vague or hazy, and I hope those that help to demarcate a "medical act" are relatively clear to the reader.

It might be said that because what constitutes a medical act involves judgements by medical professionals and given the space that has been intentionally left open for diverse accounts of moral responsibility, that the account offered here is too vague. Again, even though a criticism of the vagueness might be offered, this description of a medical act is not without virtue. It is true that it allows in a variety of conceptions, but that is in keeping with the general approach of the arguments of the text.

As noted above, one reason for the vagueness of the definition is that the parts of the medical act that are considered are those that weigh on the focus of this text: conscience. What is it that medical professionals do when then they make a judgment and how might this relate to conscience? Persons, even professionals, are whole human beings and how they interact in the world, even if within a professional context is not devoid of a variety of influences. Throughout the arguments of this text, space has been carved out for individual medical professionals to hold distinct conceptions of moral responsibility. These conceptions are understood to be part of their rich moral lives. To discount those is problematic, but if one is to do so—from with professional associations, from within medicine, from within particular communities or society more generally—the arguments needed must engage moral responsibility. This led to the recommendations in the previous chapter about public philosophy.

There are at least two ways in which we might think about medical acts in connection to conscience. First, we might think that it is reasonable for persons to hold and act according to their own conceptions of moral responsibility (at least until public philosophical debate yields a definitive answer). If this is correct, then a good deal of space should be afforded individual professionals for their conscientious practice. There is something psychologically (or perhaps anthropologically) appealing about this approach, beyond the clear answer it gives in cases of appeals to conscience. When considering medical professionals as whole human beings and thinking about the ways in which conscience, judgment, and action might be connected in this context, it is worth remarking that medical acts are performed by human physicians, not robots. That is to say that physicians can bring their critical evaluative skills, focused on the care of the patient to be sure, but that that focus could not be rooted in their whole person seems naive. It appears, as noted in previous chapters, that some CDs believe this is the case and might think of medical acts as involving merely clinical judgement. This is not to say that they do not believe that physicians have broader obligations, rather it is to say that they believe physicians do not have rich professional identities. Physicians have both rich professional and personal identities, and conceptions of moral responsibility are germane to both. Consider a physician who is asked why they scrub their hands diligently before performing a surgery or examining a patient? They respond, "It is the right thing to do." This is not as easy an answer for the CD to explain as one might think. For the CD, it appears that the reasons that might be offered in these kinds of contexts cannot both be categorized as medical and as, broadly speaking, and for lack of a better term, moral. "It is the right thing to do" should be a medical science-based answer, not a more nuanced response rooted in what would fall into a Rawlsian understanding of a comprehensive doctrine. But consider even those kinds of responses. This physician might reply, "I don't want to infect the patient" or "I won't violate a sterile field." Even these answers are value-laden, and if one has heard them uttered, they are offered a tone that suggests even if they mean something rooted in medical science they are being used as something more.

The arguments in this book suggest taking more care with distinctions like this, with taking the humanity out of judgment, with dividing the public and private, with bifurcating the selves of medical professionals. This is not to say that personal preferences of physicians should determine a course of treatment, they should not. Rather it is to say that it is disingenuous to think of medical professionals as automatons, and if it is reasonable for them to hold their own conception of moral responsibility, as it is reasonable for others to do so, then drawing from their whole person in making that determination (given the absence of a sound and definitive public philosophical argument) should be admissible. And this might, which leads us to a second way in which conscience and medical acts are connected, simply by how medical professionals understand what it is they are to do as members of that profession. This second line of reasoning brings conscience, and the need to protect it, to the fore.

In discussing conscience and the notion of integrity, Jorge Garcia writes:

> What is called integrity is not simply, or even chiefly, a self regarding virtue… Conscience protection is misconceived in much current discussion: acting from conscience in many controversial cases is merely insisting on acting in accord with (virtuously within) one's role... Being conscientious here is not, as some assume, a private, internal affair, even a kind of self-indulgence. Rather, it is a matter of keeping faith with what one has solemnly and quite publicly professed herself to do and to be for others.[4]

It is, again, how one understands oneself to be bound up with others and—for some CPs, like Garcia—what it is that one professed to do as a medical professional. The history of medicine is rife with significant violations of other human beings, including cases where communities and societies turned a blind eye or tacitly accepted the existence of such violations.[5] Whether to respect the diverse ways in which persons understand themselves to be connected and to interact with others or to leave space for those who wish to keep faith with their understanding of what they promised to do, respecting conscientious practice does not only seem defensible, it may very well be needed.

---

[4] Garcia (2020, p. 494).

[5] Many of these are well known. For a recent book on especially egregious examples within the medical research space, see Elliott (2024). Of special interest here is Elliott's analysis of whistle-blowing and the kinds of reasons that are often offered for, or what might be called instances of conscientious practice.

These ways of approaching the subject, such as respect for diversity of conceptions of moral responsibility or for professional's understandings of what they had professed (and presumably continue to profess), help to make sense of some common cases and phenomena that are often discussed in medical ethics. Consider, for example, the moral distress that sometimes arises when medical professionals are put in situations in which they feel that they must do something that they do not believe is best for the patient or that they believe is against their professed values associated with good patient care. Some of these cases are informed by professional norms, but others are informed by more general, human ones. The professional who takes additional time to explain risks to a patient after the patient has signed the consent form might be such a case. The professional who advocates for a different surrogate decision-maker for their patient, knowing that this listed person, given the surrogacy hierarchy where they practice, does not know the patient as well as another person does. Consider the professional who is motivated to act in these ways because their sibling faced a similar challenge or a friend was put in a tough spot in a similar situation years ago, or the like.

To take up a more concrete case, consider a situation in which a patient is deemed—and correctly—to be beyond the interventions that medical science has to offer. The patient's family maintains that the patient is a "full code." Distress is rising among the care team, who does not wish to perform a code because they believe that it would not aid the patient and might cause additional suffering. This is a challenging but not uncommon case and illustrates the topic well. If a full code is initiated and a professional refuses to participate, how should we think about their act? The aim of introducing this case is not to adjudicate it, but considering one response recommended in the literature will prove instructive. One might argue that medical professionals are never required to offer and follow through with non-beneficial (or futile) treatment, but that is not the issue. The challenge is how to interact with others when disagreement exists. One response to this challenging situation which made its rounds in the bioethics literature recently[6] was the suggestion that clinical ethicists ought to aid the care team in deceiving the decision-makers into making the right decision, that is, to changing the code status. The arguments for this were, unsurprisingly, broadly consequentialist, and arguably utilitarian. The outcomes are much better for all if the code status is changed. But what if a medical professional, in addition to holding the aforementioned view about non-beneficial treatment, also holds that it is wrong to deceive patients, even for good ends? If these practices were to become common, space for her conscientious practice might be needed.

[6] Meyers (2021).

5.2 Objections

With a full picture of the account on the table, it is now necessary to consider some important objections to the overall view offered in this text. In this section, a number of those objections are considered with replies to each being offered. Throughout the course of this book, objections to particular claims made and about specific arguments offered have been given responses. This section engages objections to the project, as a whole, after which, I turn to conclude the book.

5.2.1 Power and Judgement

The first objection to consider is rooted in concerns about power and the potentially unchecked supremacy of physicians, or any health professional for that matter, who has the opportunity to avail themselves of conscience protection. The language of power is often used in describing medical encounters wherein a patient is rightly understood to be vulnerable, both because they are (likely) ill and (very likely) have knowledge and ability differences as compared with the physicians, relative to the issues at hand. Discussion of conscience is a place where this language is common. Savulescu in an important and influential paper on the subject opens his examination of the subject matter describing those who would appeal to conscience protections as cowards.[7] The opening line of that article reads, "Shakespeare wrote that "Conscience is but a word cowards use, devised at first to keep the strong in awe" (*Richard III* V.iv.1.7)."[8] Another CD, whose work has figured prominently in the arguments of this text on disagreement and on religion, Stahl, describes power, invoking the term, as an issue of the first order. Stahl is quoted in a popular piece describing a panel at Yale Law School, in which her position, simply titled "Against Conscientious Objection," opens with a power-focused framing. Stahl notes, "I want to be clear that the argument I'm making is about pluralism and power."[9]

Does respect for conscientious practice, as it has been construed in this text, afford physicians with too much power? The answer is no; at least not more than they already possess. Three points help to make this case. First, if one is worried about the power differential, and with good reason, between physicians and patients, there are a host of places to address this issue which rank above allowing physicians to embrace[10] or refrain from certain practices, treatments, and approaches. Depending on the context one might argue for better language services in situations in which there is not linguistic agreement between patient and physician.[11] One might also

[7] Savulescu (2006).

[8] Savulescu (2006, p. 294).

[9] Gwizdala (2022).

[10] There do exist positive opportunities as part of conscientious practice, contrary to what some CDs argue in conceptualizing all relevant issues as instances of CO.

[11] Pilkington and Campoverde (2020).

argue for fully informed consent being required in all cases—both those that engage research and those that offer common care options.[12] Finally, one might note the vast vulnerability created by the financial machines that drive healthcare, at least in the US, and find addressing this inequity to be a much higher priority on the list of needs that address concerns of power than refusing to respect conscientious practice.

Second, there are ways in which the potentially additional power imbalance exacerbations might be slowed or nullified. If professionals are, as noted earlier in the text, required to promulgate descriptions of what their practice looks like, when they practice conscientiously (that is, the kinds of services offered or not offered, the approaches generally taken or not taken, et cetera), situations in which a patient might be surprised, confused, or unable to ascertain, and then access the kinds of care they need or request would be significantly minimized. Further, if practicing conscientiously means for a particular physician that they do not offer certain services that other physicians who are classified as similar to them offer, they will still have to let their moral responsibility claims play out in the public philosophical space—this may result in restrictions, such as might market forces, if a physician's conscientious practice limits the opportunities they have to see more patients.[13]

The third point in support of a negative response to whether respecting conscientious practice in medicine gives medical professionals too much power, focuses on the "not more than they already have" clause. The first two points highlighted ways in which conscientious practice might be understood to be less of a risk of exacerbating power differentials than other factors, suggesting that other matters should be addressed first and that ways in which such practice might be restricted so as to not allow for power differential increases could obtain. The third point focuses on the power already held by medical professionals, especially due to their judgement or the deference that is given to their expertise. I do not mean to suggest that deference should not be given to experts, but rather that the power differential in light of this is sufficiently wide that if the first point were not relevant (that is, if the more pressing and impactful items on the list were already addressed) and the suggestions housed in the second point were put in place (restricting conscientious practice or, at least, some of the potentially negative ramifications of it), it is not clear how great an impact restricting conscientious practice would have on considerations of power. To explain this claim, two points are needed: first, one about expertise; second, one about the expansiveness of medical judgment. I take up each in turn.

In their classic text on expertise, Chi, Glaser, and Farr describe a number of crucial characteristics that experts possess. There is insufficient space to survey and engage all, but two are especially relevant to this portion of the response to the objection from concerns about power. One crucial characteristic is that "Experts excel mainly

[12] For egregious violations of this requirement, see Elliott (2024). For violations in practice so egregious it is challenging to describe, see Pilkington (2020a) and Pilkington et al. (2022).

[13] This might be the case simply because they devoted significant time to each patient, given that they prize informed consent or value clear communication, however long it takes. It might also be the case because their restricted set of services means fewer patients will be coming through the door.

in their own domains."[14] A second crucial characteristic that is germane to this discussion is the requirement that "Experts have strong self-monitoring skills."[15] Coupled together, these features of experts, or of non-physician understandings of expertise, reinforce and emphasize the power differential. We understand that experts are supposed to excel in their own domain and so the physician ought to know and do the right thing, with respect to a patient and in terms of medicine. Thus, deference to *the good doctor* is not uncommon. The second feature is both more interesting and more impactful on this point of discussion. Because experts are understood to have strong self-monitoring skills, we envision them "staying in their lane" as it were and not stepping out into spaces in which they are not experts but still acting like one. However, medical professionals do not always possess this trait, and we do not always catch them when they fail to exhibit it.[16]

Consider the physician who makes claims outside of medicine, but those recommendations carry a certain rhetorical force in light of the fact that they were offered by a physician. Consider also a physician who begins to sell "health products" such as supplements in their office. Even if these health products are not approved in the ways that "medicine" must be, because they are categorized differently, they have the air of effectiveness given that a physician, by selling them implicitly endorses them. However, there are not only negative examples of physicians not staying in their lane or of espousing an understanding of that lane which is wider than might be understood. The expansiveness of medical judgements is vast. Physicians make a host of determinations about any number of patients and do so regularly. When speaking with a patient about care options, when offering recommendations, when prescribing medication—that is, when performing medical acts—physicians make decisions and do so for a variety of reasons. Physicians do not offer patients seventeen options for pain reduction after surgery, they only offer a few. Physicians, in this case and in others, artificially restrict the potential options available to patients because they believe one medication might be better than another. They do not, nor are the expected to offer every option, even if all options fit the bill. What of physicians who do their homework and learn that a patient lives in a food desert and so prescribe medication to address the patient's disease instead of recommending healthy foods because they understand these not to be readily available or easily accessible? What of physicians who integrate a variety of medical approaches (i.e., employing integrative medicine) in their practice, or those who practice holistically? These are distinct approaches, even if overlapping, and each opens up a greater possible set of responses to patient needs. However, all viable options relative to each component of care are not offered. The point here is not that physicians are constantly failing in their aims to fully inform patients as part of a consenting (or shared decision-making) process. Rather, the point is that physicians make judgments all the time which reduce the possibilities for patients. It will be said that many such judgments are about things

[14] Chi et al. (2014, p. xvii).

[15] Chi et al. (2014, p. xx).

[16] Additionally, if research on therapeutic misconception is accurate, we have a great dela to be concerned with here. See Appelbaum et al. (1987, 2012).

which are negligible—the choice of one painkiller or another does not matter—but that is still a determination that is made and made without a consultation with the patient.

Thus, it is true that the power differentials between physicians and patients are great in many ways. It is not true that respecting conscientious practice will greatly exacerbate that gulf, given the three points articulated above.

5.2.2 Negative Framings

A second objection that must be dealt with is that to respect conscience in medicine is a negating phenomenon and removes options from patients, which is problematic. This objection is wrongheaded, but merits three responses. First, it is true that respecting the conscientious practice of individual medical professionals would, likely,[17] carve out space for professionals who may opt for more limited sets of services. This notion was addressed above, but given the response to the first objection (also addressed above), there is good reason to think restrictions already exist and that respecting conscientious practice would not significantly exacerbate them. Second, though just a point of clarification, arguments in this text lead to conclusions in support of respect for the conscientious practice of medicine by individual physicians. It is true that this may include some of what is traditionally understood as conscientious objection, but that is not always the case. This is both for the conceptual reasons addressed in previous chapters, but also because conscientious practice need not be negative in the way that CO must be. This also means that the responses to conscientious practice need not be the same responses as those to CO. Consider, for example Wicclair's analysis of the issue and potential responses.He notes that full deference to CO, which he calls "conscience absolutism" can be responded to in one of two ways. First, he writes, "to counter conscience absolutism and prevent conscientious objection in medicine from running amok," one might respond with what Wicclair calls "non-toleration," which is to decline to accommodate physicians who refuse to provide legal, professionally accepted, clinically appropriate medical services within the scope of their clinical competence."[18] Notice here the stark language, as well as the clear but unnuanced response to what is, indeed, construed negatively: a refusal. A second option, according to Wicclair is "compromise or reasonable accommodation," which, "is to impose constraints on accommodation."[19] Wicclair opts for the second option, arguing that the first is "an excessively blunt instrument to prevent conscientious objection in medicine from running amok."[20] It is a blunt instrument, in part, because it is responding to a characterization that is not often as nuanced as

[17] Though this may depend on the particular accounts of moral responsibility offered.

[18] Wicclair (2019).

[19] Wicclair (2019).

[20] Wicclair (2019).

it might be. Refusals, conscientious objections, and the like are not each coextensive with the broader and richer notion of conscientious practice.

The third point is related to the second and it is that conscientious practice encompasses opportunities for positive exhibitions of conscience, not merely refusals.[21] Conscientious practice need not always include or be focused on opting out or restricting options. It includes positive medical acts, that is, instances of expanding medical services. This can be done in ways which need further examination, such as medical professionals selling health products, as noted above, but it can also highlight the kind of integrity that medical professionals (often) exhibit. Recall the description of integrity in relation to conscience referenced earlier in this chapter: "Being conscientious here is not, as some assume, a private, internal affair, even a kind of self- indulgence. Rather, it is a matter of keeping faith with what one has solemnly and quite publicly professed herself to do and to be for others."[22] Thus the physician who cares for transpersons in a state in the US, for example, which is hostile toward transpersons is a kind of practice that ought to be protected. It may be that specific procedures, such as gender affirmation surgeries are not reimbursed or not supported in a particular state, but if the arguments of this book are sound, such space should be carved out and the physician ought to be protected in caring for such patients and offering such treatments. Similar kinds of cases might include providing abortion services in the U.S. post Dobbs,[23] allowing one to put to the side interstate travel and related considerations.[24] Similarly, the arguments in this book would support those whose conscientious practice includes medical care that incorporated careful attention to climate change and its impact on health,[25] even in locations where this is not in accord with common cultural attitudes or where states (and insurance reimbursement policies) would not be supportive.

It should be noted that the arguments of this text support the carving out of space for conscientious practice, but they do not critically assess particular practices or approaches that might fall under the protection that is argued for. Thus, it is possible (given the limits of philosophical argumentation and, more so, public philosophical argumentation), that the opposite occurs, as well. Healthcare procedures especially relevant for the care of transpersons may be accepted and supported in a particular locale, and the arguments of this text could be used by medical professionals practicing in those areas and not offering such services. This is possible not only for

[21] This is not a new idea, but one that CDs have not given justice to. For more on positive claims of conscience, as it is discussed in the literature, see Miller (2021).

[22] Garcia (2020, p. 494).

[23] For a bioethical discussion of relevant issues, see Ziegler (2022).

[24] For more on travel post Dobbs, see Cohen (2022).

[25] For an argument that climate change is the most important determinant of health, see, for example: Ziegler and Muchira (2023).

procedures especially relevant to the care of transpersons,[26] but also to the incorporation of climate change into health assessment, care, and treatment, as well as abortion services. The point here is that respecting conscientious practice would mean, in part, supporting positive opportunities for care. Notice that is not what would come out of the deliberations of professional associations, which are, by nature, much less nimble. That said, there is one boundary that has not been sufficiently discussed in this section and exists in addition to the restrictions that might be put in place by public philosophical arguments: the medical act.

To consider how our focus on the medical act might impose restrictions, consider the following situation, which raises the question of when, if ever, a medical act can involve political claims. To sketch this case fully, as an adequate test of potential limits on conscience, given bounds placed by the notion of a medical act, the sketch must begin at a foundational level. Physicians heal; they aim at and strive for the well-being of their patients, conscientiously attending to the care of the person in front of them and, in many instances, to that person's family, friends, and loved ones. However, as has been illustrated throughout this text, this essential goal of medicine can be realized in different ways and the goal itself is understood in different ways. Respect for conscientious practice is not limitless. It applies to medical acts, as argued above, and so this leads to the question, "What are those limits?" In addressing this question, consider a thought experiment about politics in medicine. Discussion of politics has, for some time, if not always, been anathema within the clinic, and so it is a fitting test as potential limit case for the medical act's bounding of conscience. Do the outer bounds of conscientiously practiced medical care include conversations with patients about political matters? Are such conversations appropriate or not?

Significant changes in the culture of medicine and the ethics of healthcare practices have occurred since oft-cited Hippocratic origins guiding the norms of the medical act, both its science and its art.[27] This is especially noticeable over the last century and a half, and especially for physicians.[28] However, what has remained constant, even if not fully realized,[29] is medicine's primary objective: care for the patient, an objective that is agreed upon by proponents on all sides of the debate over conscience. This helps to explain the tension that some physicians feel when asked to think about medical practices aimed at the public health and medical practices aimed directly to benefit the patient in front of them. Navigating these divergent aims and doing so in diverse ways is something for which conscientious practice would

[26] An extremely troubling, but conceptually possible outcome from this approach is the use of these arguments to support discrimination. A distinction is often made in the literature about appeals of conscience being applied only to procedures, not to persons; thus, it might be ethically acceptable for a surgeon not to perform gender affirmation surgery, but it would not be acceptable not to treat transpersons. For a discussion of how this kind of distinction might fail, see Brummett and Campo-Engelstein (2021).

[27] For the best discussion of the oath and its ethical considerations, see Cavanaugh (2017).

[28] Starr (1982).

[29] Various situations support this concern, two of which include concerns about racism and sexism in medicine. For two examples of these situations, see Smith (2019) and Rabinowitz (2018), respectively.

allow space. This is not to say that these aims are always in tension; they are not. This claim is well illustrated by some behavior in the recent worldwide pandemic. Consider the heroic responses of many physicians during the heights of the COVID-19 pandemic. The additional hours, risks to self and family, and public advocacy for proper responses are all examples where care for one's patients and care for the broader public align. Consider also the relatively mundane example, which has become an enduring symbol of bioethical discussions during the time COVID-19: wearing a mask. Mask-wearing contributes to the public good,[30] but it also benefits the mask-wearer, both directly and indirectly. There is a small decrease in the risk of contracting COVID-19 if one dons a mask (the direct benefit), but the far greater (and indirect) benefit of general mask-wearing is (coupled with good hygiene and distancing practices) the reduction the spread of the virus and, thus, the decreased risk of contracting the virus.

This analysis is relatively uncontroversial, at least within medical and scientific communities, and so it helps to demarcate the limits that reflection on the medical act might place on conscientious practice. Learning of a physician's recommendation to their patient to wear a mask, physically distance, wash their hands, et cetera, may strike the reader of this book not only as uneventful, but as a common instance of good medical practice. But what of other recommendations that are also beneficial to the health of a patient? What of a physician who recommends that a patient vote in a particular way or write her political representative toward a stated end? Is this a permitted instance of conscientious practice or is it out of bounds, so to speak? It is this question, which resides at the nexus of patient advocacy and patient care, to which we now turn.

In the current or, at least, recent socio-political moment(s), physicians may find themselves faced with situations where the means by which they might best care for their patients, given their own understandings of how they are to interact with them, could include work that might appear outside of the traditional medical realm. Whether accounting for implicit bias, being cognizant of structural racism and sexism in the societies in which they work, or treating their patients in light of an understanding of the determinants of health, the limits of care have and do evolve. Are there acts which are instances of advocacy and also properly described as medical acts? Do acts (or the series of actions) that yield the starting of anti-sexism groups within a hospital that one works, or writing to political leaders for additional resources for patient care, or marching to protest racial injustice count as medical acts? Fully evaluating this important class of possibilities is beyond the scope of this text; however, let us consider one instance which carries the trappings of a medical act, as it might occur in a physician's office, during a yearly checkup, and amid a host of other medical acts.

Consider a physician who seeks out information about where their patient lives and tries to locate healthier sources of food for their patient. Suppose they are successful and locate a place, albeit a bit out of the way, and share and strongly recommend that

---

[30] Ueki et al. (2020). For a discussion of best practices applied to institutions of higher education, see Pilkington et al. (2021).

their patient seek out this kind of food. Is this physician practicing medicine (i.e., do the preceding sentences describe a series of medical acts?) or are they engaged in broader socially responsible practices to combat food deserts? What if the physician writes to their local representatives or national leaders to put in place policies that would reduce the number of food desserts in the areas in which their patients live? Is this medicine, advocacy, or a bit of both? This physician is advocating for particular social and political reforms which they understand will benefit the health of their patients. Are these medical acts? (The answer to this question matters both because it highlights the increasing breadth of medicine but also because requiring practices within the scope of medicine is different from cheering or tolerating the supererogatory advocacy of some who are members of the medical profession.) In doing so, suppose the physician in our example has a particular patient in mind—not just an abstract goal—one they have just seen who is diabetic and is having a very tough time finding affordable, desperately needed, healthy food. The physician is seeking the well-being of the patient, the primary aim of their profession. With this aim, they might advocate for increased government and community resources for vaccine development, in addition to their efforts to combat food deserts. They might also advocate for policing and criminal justice reforms in light of the negative health effects that unjust policing has on the patients they treat, especially their patients who are patients of color.[31] If a physician serves as a change agent on these socio-political issues because they are motivated by the health needs of their patients, are they undertaking a medical act?

To aid in answering this question, we might distinguish between two kinds of advocacy, those internal and external to their profession.[32] By *external professional advocacy*, I mean acting for the good of another rooted in a professional's membership in a broader community, for example, as a citizen or human being. By internal professional advocacy, I mean acting for the good of another rooted in a professional's practice or her identity as a member of a particular profession. For example, the recent reports of unwanted hysterectomies taking place in an ICE Facility in the state of Georgia, in the U.S.,[33] present physicians with opportunities for both internal and external professional advocacy. A physician might, as a member of the broad American community or simply as a human being, oppose such practices and voice their support for vulnerable members of the human community by contacting political leaders, raising awareness, marching in protest, et cetera. They might also engage in internal professional advocacy, that is, in light of their expertise and membership in the medical community, they might speak out against such abuses, request that professional organizations of which they are a part take a stand on such practices, seek to censure or expel from the profession offending parties, et cetera.

---

[31] See the recent AMA statement on this matter: Harris and Ehrnfeld (2020).

[32] Pilkington (2020b).

[33] "ICE whistleblower: Nurse alleges 'hysterectomies on immigrant women in US,'" BBC News, September 15, 2020, https://www.bbc.com/news/world-us-canada-54160638.

Physicians have a particular duty to regulate the practices of their profession,[34] as noted in the preceding chapters, so situations like what occurred in Georgia have straightforward, ethically supported, options for response in terms of internal professional advocacy. They also, due to the gravity of such violations of human dignity,[35] have straightforward—even if complex—ethically supported options for external professional advocacy.

Situations which involve a physician's own practice are more complicated because these situations can blur the line between these two kinds of advocacy and, as a result of this blurring, can raise questions about the appropriateness of practices proper to one category of advocacy being employed in a situation which calls for another form of advocacy. At the heart of this distinction is the question of what constitutes a medical act. To illustrate this, consider again that case type which is understood to be anathema in medicine. Suppose a patient mentions a political topic during a routine visit. They ask their physician what they think about a local race for mayor or about an election for a senator or governor? Political topics such as these have been described as taboo or, at least, discussion points to avoid.[36] However, the kind of conversation is relevant to assessing its appropriateness, and such conversations might be best governed by considerations of etiquette, shared goals, or optimal care. The first category describes superficial but potentially important category of "small talk." A physician may avoid "politics" as a topic of small talk given the potentially controversial nature of this subject matter. The second category refers to slightly weightier considerations, where political conversations might introduce obstacles in communication and thus cause challenges for shared decision-making. A physician may avoid discussing a recent mayoral race, given the divisiveness of the race and concerns that were they and the patient of different minds on the subject, the patient may be less communicative. The first and second categories are a matter of degree; they deal with comfort and effective communication as part of the physician-patient relationship. The third category is of an entirely different sort, as it has to do with a potential recommendation from a physician for the health of their patient via a non-medical means—would this constitute a medical act?

Returning to our example, suppose that in the political race under discussion, the candidate's platforms diverge significantly regarding resources to be devoted to healthcare and the physician in our example believes that it would be best for their patient's health if one candidate were to win the election and have their opportunity to enact certain policies as opposed to the other. If the physician raises this topic, it is neither a point of small talk—bound by rules of etiquette—nor is it a way to aid better communication; their aim would be to offer optimal care for their patient, which they believe requires the enactment of a particular policy. Another way of describing the physician's aim is that they aim to make their patient aware of a potential benefit to their patient which, to effectuate, might need the contribution of their patient. In

[34] Buchanan (2009a).

[35] Pilkington (2020a).

[36] Vogel (2018).

some ways, this is no different from an activity or diet plan which requires patient participation.

It is worth highlighting that this is not a case of a physician merely being asked for a personal opinion about a piece of local news. It is a case where a physician is considering recommending an action which will directly benefit the health of the patient: "Don't smoke, don't eat too many fatty foods, and don't vote for the candidate who will negatively impact your healthcare." Is this non-medical, health-focused recommendation within the physician's purview or in so recommending do they overstep?

What complicates the analysis of this type of situation, as has been noted throughout this text's considerations of conscience, is the vulnerable state in which patients present themselves and the status and power differentials that exist within the physician-patient relationship.[37] A physician should remain cognizant of these factors in this situation in the same way that they are during all of her patient encounters. If they are cognizant of these imbalances and they are clearly seeking the benefit of their patient, should they make political recommendations? Does a physician who, aware of the socio-political realities and with their patient's well-being in mind, step outside of the bounds of medicine in making a political recommendation?[38]

Categorizing this case as an overstep embraces a more traditional approach to questions about what falls outside of "the medical act." This approach avoids concerns of undue influence, of encroaching on other areas of expertise, of paternalism, and—most importantly for our purposes here—underscores that there indeed are boundaries on conscientious practice, given what counts as a medical act. However, were one to categorize this case as a medical act, this might be because squaring this approach with an appreciation for the broader structural features that impact a patient's health can be challenging. Returning to the original example that motivated this discussion, suppose in that aforementioned local mayoral race, one candidate has a robust plan to address food insecurity and the other has no interest in addressing it. If a physician votes for the former candidate because they are concerned about the wellbeing of their patient and writes local leaders advocating for better food options, why should they stop short of engaging their patient directly, one might wonder? Surely political platforms are more complicated and nuanced than this situation and a single election is not necessarily a clear determiner of whether a food desert might be watered; however, in some instances, the benefits to a patient's well-being are clear. Politics in medicine is a limit case; though reasonable persons can disagree, it does not constitute a medical act, and so, would be outside the bounds of conscientious practice. The point of the long treatment of this case is to highlight that there are conceptual limits placed on conscientious practice.[39]

---

[37] Nimmon and Stenfors-Hayes (2016).

[38] Put to the side are the practical concerns that in so recommending, the physicians risks alienating their patient or potentially jeopardizing a good relationship. These are important considerations in practice, but the focus of the argument at this juncture is whether medical acts restrict the potentially expansive claims of conscientious practice.

[39] There may also exist—practically speaking—both legal and institutional features which help to define a medical act and, thus, help to place bounds on conscientious practice. Two points are

5.2.3 Distinct Epistemologies

A third objection to the overall argument of this text is that in carving out space for conscientious practice, too much space is opened up and the diversity of the medical profession will lead to a lack of coherence and an inability to understand what that practice (or set of practices) is. That lack of coherence and clarity will, in turn, create challenges for patients in receiving similar care options regardless of where, when, and from whom they seek care.

This is a strong objection to the project. The limits placed on conscientious practice are the bounds of medical acts, but as this would not restrict diverse accounts of moral responsibility, it should also not restrict diverse approaches in epistemology. This is one objection to which I simply bite the bullet, as is said. There exist radically different theories of epistemology and diverse ways of knowing. My account engages medical practice that is steeped in evidence-based medicine, as formed by work in the traditionally understood medical sciences. However, my account would not rule out more holistic approaches to medicine, as noted above, nor would it rule out integrative physicians, what is sometimes called eastern medicine,[40] or a variety of indigenous ways of knowing applied to health considerations.[41]

5.2.4 What Physicians Owe Society

A fourth important objection to the overall project relies on the idea that there exists a social contract between medicine (or members of the medical profession) and the public (or members of society, at large). This argument might take various forms and some of these forms are implicit in objections that have arisen and have been

---

worth noting here. The claims of this text are philosophical in some cases, moral in others, but in no place are the claims legal in nature. That is, conscientious practice may differ from the legal requirements in certain jurisdictions, whether this is due to instances of positive conscientious practice in offering gender-affirming care, for example, or negatively, in requesting not to be involved in MAID. Conscientious practice is an ethical claim, not a legal one. Second, though institutions have a great deal of impact on the conscientious practice opportunities of individual professionals, institutions themselves ought not to be thought of as the kind of entities that can conscientiously practice, that is, institutions do not have consciences. Institutions might have mission statements or vision statements, or articulated institutional values, but they do not possess a conscience. There is not space in this text to take up this idea further, but it worth noting that it runs contrary to the current Health and Human Services guidelines, stating, "Several provisions also protect the conscience rights of certain federally funded health care entities to not participate in actions they find religiously or morally objectionable." (for more on this, see Rights (OCR), Office for Civil. "Conscience Protections." Text, October 14, 2010. https://www.hhs.gov/conscience/conscience-protections/index.html.

[40] Consider in this category work in ancient Chinese medicine and ethics. For one example of a comparative study of this kind of ethic and more common, four principles, approaches in bioethics, see Tsai (1999).

[41] Consider also work in the human rights space on this subject, for example, see Muyskens (2024).

offered responses in previous chapters. However, so as to not give short shift to this important, overall objection to the project, I take it up again here, with all of the conceptual machinery of the account on the table and with a focus on one of the more compelling ideas from CDs: reciprocity.

What medicine might owe society and what society might owe medicine are important topics to tackle, but the parenthetical formulations of these ideas are more germane to the arguments of this text. There exist longstanding questions about the relationship between members of the medical profession and members of societies in which physicians live and work. As noted above in the discussion of professions, the hallmark of a profession is that it exists to be in the service of others. If this altruistic requirement is not satisfied the group in question is not a profession. Though altruism is a necessary feature of professions, as noted above, it is not sufficient condition for the existence of a profession.[42] Borrowing from Buchanan, it was argued above that professions are social constructions, not facts of nature and, therefore, must be justified. One straightforward justification for the privileges given to members of the medical profession is the service they offer in caring for members of society and the great need that exists for that service. Such privileges can include a variety of external goods: excellent salaries or preferred parking spots, and also a respected status in society. If members of the medical profession stray from their service orientation and other members of the profession cannot curtail these behaviors, regulatory bodies enter the picture, which brings us to the question at the heart of this objection: does respect for conscientious practice lead to a failure of physicians to keep up their end of the social bargain?

The answer is no. It is true that physicians must be altruistic, that their work must be service oriented in order for them to be counted among the number of "professionals" in existence, but how this service is to be realized can take a variety of forms. One need not have violated one's contract with society because one has a different conception of moral responsibility or approaches health in a holistic or more particularized manner or because one positively exerts one's conscience by offering care (or negatively by restricting one's own service output) that is not generally accepted in a particular locale. As altruism can be realized in numerous ways, this objection misses the mark.[43]

5.2.5 *Artificial Conscience?*

A final objection to the project serves two purposes; first, it engages an important set of considerations which are critical of the work in this text; second, it engages the

---

[42] Buchanan (2009b).

[43] Physicians should always care for those in need. There are cases where doing so might be a test of their empathy or resilience. However, this is not a one-way street; members of society should also think about this contract and wonder, as many health professionals are leaving their fields or retiring early, whether they are doing enough to uphold the other—and less discussed—end of the bargain.

topic of artificial intelligence (AI) in medicine, and in light of the increasing (and potentially exponentially increasing) impact that AI is having on healthcare, no book written at this time can avoid addressing it.

This final objection is unlike the others. Were the others successful, issues with my accounts—to varying degrees—would have to be dealt with, some of which were potentially devasting to the view. Objections to my approach to conscience, related to AI, are distinct; they pose equally problematic challenges—if they are successful—but the result of these challenges is not that respecting appeals to conscience in medicine is wrongheaded or that such respect fails to appropriately respond to CDs' views. Rather, in one place, it is that resolving disagreement becomes much more challenging, as a new kind of interlocutor or partner is in play. In the second place, it is that the erosion of the medical act—as AI does not act in the same way, nor would collaborations with AI yield similar assessments of an act—has begun, and a new focus on conscience in medicine would be needed. In the remainder of this section, I set the stage for this discussion, then turn to the objections.

Amid the ever-increasing literature on the role of AI and its implications for medical professionals and ethics, one set of questions germane to the discussions of this text has received surprisingly less attention than other sets of questions: those bound up with considerations of the profession on medicine and, derivatively, with conscientious practice. This is surprising for a number of reasons, foremost of which is that the increasing inclusion of AI in healthcare is—at least for certain kinds of systems[44]—an incursion into the spaces in which professional judgement dominates or has traditionally been understood to be dominant. Medical decision-making, for example, which had been solely the purview of the physician has (and for the better) become the purview of patients *and* the clinicians who care for them. Yet wherever one pauses along the spectrum (both theoretically and historically) of complete physician paternalism to perfectly orchestrated shared decision making between patients (and loved ones) and a care team, the purported root of ethical reflection rests (at least in some ways) on the relationship between patient and medical professional, that is, between a (or multiple) member(s) of (a) care team and a patient. If a third party joins in to aid in care, what are the implications for conscientious practice of professions and the implications thereof, given that the third party in this case is not a member of a profession?[45]

As the power and influence of AI expands within healthcare, the potential for benefiting health outcomes is significant, but insufficient work has been done to properly conceptualize the role of this technology with respect to the conscience of professionals. Through a professionalism lens, this raises the question, "What is the

---

[44] AI is used here as a catch all term, but it is important to note that there are a variety of algorithms that might be employed in a variety of different manners and places. They are distinct and so are the relevant professional and ethical impacts. An algorithm created to aid in hospital intake processes might not raise the same ethical issues as clinical decision support algorithms, with resulting implications for conscience and professionalism being distinct, as well.

[45] One might reject this analysis, citing that business and other financially focused interests have impacted medicine for some time. This is correct, but misses the point, as the issues under consideration have to do with medical acts and with disagreement within professions.

professional status of particular AI systems: are they 'colleagues' to be worked with or tools to be used?" The answer to the question bears on topics at the heart of this book: moral responsibility, conscience, and disagreement.

Discussions of "AI" in healthcare can sometimes be misleading, as there are a variety of different kinds of artificial intelligence and depending on the particular system's function and abilities, the answer to our question might change. Consider one specialty in medicine, surgery. Surely there are AI systems which are tools, just as a scalpel or a suture is a tool. Even if we ratchet up the degree of complexity of the object—we might discuss laparoscopic cameras or even robotic surgery machines, such as the Da Vinci Si—we describe situations involving these items as cases of professionals operating machines. Consider some common phrases: "It was useful for the surgeon to have that third robotic arm in that case" or "This surgeon is particularly adept in employing those sutures" or "Which scalpel did the scrub tech prep for the physician to use?" However, there may be other AI systems which are not merely a more complex robot but fit better the description of a professional colleague—those making the kinds of medical decisions that professionals used to make, those collaborating in a significant way with those professionals in making decisions, or those employed to run (or *lead?*) other algorithms and systems. In the instance where an AI system functions to assist in a healthcare professional's decision-making, the AI system, properly understood, appears to be acting as a professional colleague. If that is the correct assessment, the implications are significant. Answers to these questions about AI in healthcare have implications that bear on the relationship between medical professional and patient, the privacy of patient information, and the role of informed consent in therapeutic and palliative procedures, and many others. They will greatly impact how healthcare is delivered in the years to come and the human professionals who deliver that care.[46] Further, if AI systems are to be considered "colleagues," then the norms of the profession apply to those systems and their creation and programing must internalize such norms; however, if AI systems are merely tools for the use of professionals, then they need not be built in accord with professional norms, as the medical professional is where that onus resides. In some ways, it is a question of who acts—the physician or the AI?

Conceiving of particular AI systems as colleagues may seem easier or more challenging depending on the details of one's particular account of professions and, derivatively, of professionalism. If one prizes the professional duty to place the patient first or to avoid discrimination,[47] these might be duties that can be written into an algorithm and, in fact, done so in such a way that AI is less biased than its human counterparts.[48] To the contrary, if one prizes openness and honesty with patients, then AI might be incapable (at least at the time of this writing) of being a

[46] For a critical discussion of AI decision support in terms of ethics-specific issues, see Pilkington and Binkley (2022).

[47] For one example of a discussion of discrimination and AI, see Binkley et al. (2022).

[48] For a discussion of AI and bias, see Binkley et al. (2022).

professional since it cannot explain (to a patient or even to another medical professional) why particular decisions were made in particular cases, what AI experts often refer to as the explainability problem.[49]

Consider a situation in which a clinical decision support system is employed in a surgical case, and we might ask, is the surgeon the professional with an opportunity to practice conscientiously or is the algorithm the professional, performing the medical act. Before assessing this type of case, a discussion of professionalism is needed.[50] Though there are a variety of accounts of professionalism, ranging from lists of virtues to prohibitions on certain types of behaviors to the unfortunate but common assessment "We can't tell you what it is, but we know it when we see it," let us focus on one helpful account, that is both concrete and specific to the situation (clinical decision support in surgery). In their work on the subject, Christian, et al.[51] follow a fairly common description type rooted in the idea of a social contract between professions (or members of professions) and society,[52] writing, "In essence, the basics of professionalism are quite easy to articulate. In return for professional autonomy, self-regulation and a recognition of their unique place in society, the public demands of physicians accountability, ethical standards and an altruistic manner of delivering care."[53] As discussed above, and with many accounts of professionalism, these physicians prize a service or altruistic focus, continuing, "Altruism, or the putting of the interest of patients and society consistently ahead of one's own, is the bedrock of professionalism for the physician and has been recognized as a key, unifying concept by several professional bodies. Altruism has also been defined as going "above and beyond" one's call of duty."[54] Thus, for these medical professionals, in their attempt to articulate an account of professionalism for the Canadian Association of General Surgeons, eight key features can be identified:

1. Duty to consider first the well-being of the patient
2. Respect for patient and clinical autonomy and providing the highest quality of care
3. The adoption of new technology, partnership with industry and participation in research that benefits the patient
4. Care without discrimination
5. Working with other health care professionals as a team for the benefit of patients
6. Openness and honesty with the patient and disclosure of adverse events
7. Accountability to the courts, licensing bodies, peers and hospitals
8. Balance between professional and private life

---

[49] For a discussion of this problem in connection with healthcare, see Binkley and Green (2021).

[50] For a discussion of relevant ethical issues related to clinical decision support, see Binkley and Green (2021).

[51] Christian et al. (2008).

[52] See the aforementioned objection and discussion for considerations of conscience relative to social contract theory and medical acts.

[53] Christian et al. (2008).

[54] Christian et al. (2008).

For the purposes of this argument, let us accept these eight features as encapsulating what it is for surgeons to embody professionalism in their practice. With this in mind, let us focus on clinical decision support and professionalism. One place in which professionalism can notably be displayed, and one that is important to this text, is in situations of disagreement. In fact, given the increasing emphasis on teamwork in medicine, it might be that instances of disagreement are among the most helpful models for testing the existence of professionalism.[55]

Before attempting to resolve such disagreement, we must first understand the professional status of a clinical decision support system, and the potential divergence between the values of the medical professional and those of the latest member of the team, the clinical decision support system. This introduction complicates attempts to resolve disagreement because of potential divergences between a surgeon and the clinical decision support system are not like disagreements over straightforward, even if nuanced, intraoperative decisions between two surgeons.[56] I now turn to explain why that is the case, what specific issues arise from this difference, and what the potential responses to those issues are.

A disagreement between two surgeons about an intraoperative decision has clear, even if potentially contentious, paths to resolution.[57] Suppose Dr. S and Dr. R disagree about whether to x or to y as they operate on a patient. Due to the time demands of the surgical intervention, x may be chosen because it is the choice of Dr. S, who is the more seasoned surgeon or because it is their patient and Dr. R is merely assisting on a complicated case, or because Dr. S is the surgeon of record and Dr. R is the assistant. In each of these instances, a procedure is in place to move beyond a disagreement based on reasons that are accessible to all: deference to experience, recognition of the special relationship between a particular surgeon and their patient (and ensuing obligations), or stated hierarchies of the roles of each surgeon in a particular case. A fourth, and superior option, would be if the disagreement were resolved because Dr. S shares with Dr. R something about the surgery which was not noticed by Dr. R. Dr. S explains their reasoning to Dr. R, who after evaluating what is shared and visualized, agrees with Dr. S.

---

[55] The potential for AI decisional (and active) support is an exciting technological advancement which may aid surgeons, operating teams, health systems, and—what offers the appropriate and ultimate justification in these settings—the patient. It has been argued that ethical reflection on emerging medical technologies often lags behind the existence and use of those technologies (and we believe this is the case for professionalism-based reflection, as well), and proactively considering future possibilities is essential to ethical surgical practice. This is not an isolated claim, as the COVID-19 pandemic made clear: healthcare institutions have a duty to plan for future eventualities, and so surgeons may face greater pressure to consider, evaluate, and respond more quickly to AI introductions in the surgical theater than prior to the pandemic.

[56] For instance, disagreements between surgeons and clinical decision support systems could be rooted in values differences relative to the data that a system is trained on, the values of particular surgeons (in terms of the space carved out for their conscientious practice, or either's "understanding" of the values expressed by the patient).

[57] It is reasonable to assume that they share the same, or somewhat similar values (at least as contrasted with the multitude of values that might be encoded into clinical decision support systems, such as care not only for the patient, but for public health, or a health system's financial health).

In a way, these considerations are the challenges of sharing agency or sharing accountability. Dr. S and Dr. R can communicate with each other, explain their reasoning and what they are seeing, and move forward. This is not to say that they will always agree, but when they do not there are at least three possibilities in place for moving forward. This is not how interactions between Dr. S and a clinical decision support system would progress. This is because clinical decision support cannot explain itself to Dr. S. If the clinical decision support system recommends *y* and Dr. S is of a different mind, how should Dr. S proceed?[58]

Instances of disagreement can often prove illustrative of motivating principles, background assumptions, and deeply held values which parties bring to a situation of conflict. Thus, exploring conflicts and potential resolutions might shed light on what it would mean for the medical act and the role of conscience to have an AI system involved. Conflicts between the surgeon and the AI system might be rooted in, first, differing priorities. The surgeon may prioritize the health of the patient or the integrity of particular structures of the patient's body which would allow the patient to perform daily tasks that are important to them. The AI system, while aiming at these goods, might prioritize more highly the cost of the surgical intervention or of the hospital stay that is statistically associated with surgeries that exceed a specified amount of time, for example.[59] Second, they might be rooted in the lack of clarity of the relationship between a surgeon and the AI system: is the clinical decision support system or the surgeon the decision-maker? If the surgeon defers to the clinical decision support system, have they become more like a flesh a blood da Vinci Xi and the AI like today's surgeon performing a robotic surgery?[60] Third, a related but distinct issue is that of ownership. Even in the most interprofessional of teams, there is someone who has the final say, someone with whom the covenant that informed consent relies on, someone who speaks with the patient after the surgery. Who is to take ownership over the intervention and to satisfy the duty to care for the patient in the case of an AI assist? Answers to these questions and concerns speak to who (or what) is performing the medical act and how this might impact claims of conscience in medicine.

Each of the aforementioned conflicts arises due to the inability of the clinical decision support system to explain itself to the surgeon. This matters because if a surgeon relies on a clinical decision support system but does not understand its

[58] The difference in these cases illustrates the problem of explainability. Explainability is a necessary hurdle to jump before this human-AI surgical team should operate on patients. It is not an impassable obstacle, but it is also not easily overcome, and it is at the root of three issues of the potential values differences.

[59] These are simply examples and could give rise to the same conflict if the assigned priorities were reversed with the AI prizing the integrity of particular structures and the physician expressing concern over financial burden, et cetera.

[60] This lack of clarity—connected to issues of shared agency—raises concerns for an ethical consenting process and for the shared decision-making process in medicine, which is at the heart of the relationship between a surgeon and a patient, and at the forefront of a surgeon's professionalism. (This is especially the case in regard to qualifications 1 and 6 in the preceding account of a profession, and—more broadly—with the altruistic focus of professionalism described by Christian, et al.) For more on shared agency, see Binkley and Pilkington (2023).

decision process, it raises questions about whether the surgeon can properly hold the accountability for the action being performed.[61] Were explanations offered in these cases, the absence of a conflict resolution procedure keeps the problem in place. If the kinds of resolutions suggested above are employed, a great deal of medical professionalism needs to be rethought for the AI's decision to be accepted; the implications for conscientious practice and understandings for the actions medical professionals undertake are significant.

These are some of the key issues at the heart of what the categorization of AI in healthcare means for conscientious practice. If the introduction of a clinical decision support system (and related AI systems) means that it (they) is (are) more properly understood as the agent than the physician, who is performing medical acts in these cases, and what space for conscientious practice should—*if any*—be argued for? If the introduction of clinical decision support does not lead to a situation in which the physician is replaced as the agent, then how is potential disagreement resolved?[62] These issues are complex; even the discussion above oversimplifies the differences between AI-human and human-human medical professional relationships. Compare the situations. Dr. S and Dr. R, even if they were trained in different locales, grew up in different trends in surgery, manage patient care more or less conservatively than the other, etc.; they each have *professed* to do and to be something, a something which entails certain obligations to patients, to colleagues, to future colleagues, and to members of the societies in which they are a part. They share a profession; they are bound by the Hippocratic Oath; they aim—*arguably* above all else—to benefit the patient, at least as they understand the proper bounds of medicine? What would it mean for a clinical decision support system to make such a profession? And does it matter that it cannot? This highlights the depth of the chasm between human surgeon and clinical decision support systems. As the philosopher Alasdair MacIntyre notes in his work on rival accounts of justice, the challenge is not merely deciding between two accounts of rationally "[f]or those rival accounts of justice are each framed in terms dictated by, and in some respects not detachable from, an overall system of thought, one which both imposes constraints upon its philosophical parts and also provides standards by appeal to which each philosopher advances a rational justification for his particular account."[63] In this way, surgeons are of one philosophical school, being able to argue, adjudicate, and collaborate in light of their participation in a shared goal and set of practices; AI, as far as it is unable to offer an

[61] An objector to this claim might suggest that as technology advances (after a clinical decision support system is placed into practice and the algorithms are appropriately trained), the particular intraoperative decisions it makes will be explainable. In response, consider that the kind of explanation needed may be more complicated to achieve than it might appear at the outset—and this is to put to the side the ethical questions about training the AI and the transition period between current practice and this future state. (For an example of this discussion, see McCradden et al. [2022].)

[62] For a discussion of attempts to resolve disagreements with ethics and AI, see Pilkington and Binkley (2022).

[63] Macintyre (1991).

explanation of its decision to the surgeon or, if it could, to make the kind of profession that binds it socially and collegially, operates in another system of thought. However, unlike those in the debate over conscience, it cannot even offer reasons for others to consider.

5.3 Conclusion: Living with Moral Disagreement

An early draft of this book concluded with a final section titled "Living Well with Moral Disagreement." The reader will note that this is not the current title of the book's conclusion. Living well with moral disagreement may have been too ambitious an aim, and for at least two reasons. First to have a claim to contributing to that lofty conversation, a book would have to offer more in terms of both breadth and scope. This book will be successful if the reader recognizes that disagreement exists within the healthcare contexts, finds the roles that physicians play to be more complicated—both in terms of their approaches to practice and also their accounts of moral responsibility—than originally thought in what it is physicians are doing when they act medically. Living with disagreement in medicine and understanding what contributes to it—not religious claims, at least not in the way that they are commonly understood to be problematic—within the conscience space is a sufficiently satisfactory goal, especially given the deeply entrenched positions and (sometimes) shrill arguments in debates over conscience in medicine.

This book is an attempt to take seriously the existence of moral disagreement and the claim that reasonable persons can disagree. Having laid out a variety of positions on the role of conscience in medicine, I described two primary clusters of positions—one in favor of protecting conscience and one against. In the course of critically engaging the debate, the text spends significant space addressing disagreement and an account of a profession. Relying on this work, it rejects conscience dissenter positions that rely on professions and professional norms (but really mean the norms of professional associations, and given that different professional associations have different foci, approaches, and detailed levels of concern, probably mean one particular professional association—though no viable candidate was found) to guide the acceptability of conscientious practice and to explain and adjudicate disagreement. This led to what is arguably the most controversial claim of the text, that the debate over conscience in medicine is not about religion, but that it is actually about differing conceptions of moral responsibility. This led to a discussion of moral responsibility and a call for public philosophy. The text concluded, penultimately, with a brief discussion of medical acts and, finally, with replies to a series of serious and important objections to the arguments of the text.

If the arguments included in this book are sound, I hope to have moved the conversation about conscience away from the overtrodden and deeply entrenched debates over religious conscientious objection in medicine, and opened up space for potential agreement but also for a wider understanding of what respect for conscience might mean.

The work is not complete. If these arguments are sound, then physicians in training need additional educational opportunities to hone critical skills, given what lies on the horizon. A more robust and philosophically defensible account of an act and a particular candidate for an account (or sets of reasonable accounts) of moral responsibility could be offered to help to address concrete cases; that is another project, but it is important to wade into the public philosophical debate, now that the groundwork has been laid, to see if there are such conceptions of moral responsibility that are indefensible in medicine, which ones are defensible, and what those answers mean for the protection of conscientious practice generally. Penultimately, much more work is needed on the professional identity of medical professionals in order to do justice to descriptions in this text of professional identity, of self-conceptions, and ultimately of a practical, physician-focused perspective of the medical act. Finally, the terms "physician" and "medical professional" were used (and though it is true that in many instances these terms could serve as placeholders for any health professional), and more work is needed to extend these arguments to members of other health professions.

References

Appelbaum, P. S., L. H. Roth, C. W. Lidz, P. Benson, and W. Winslade. 1987. "False Hopes and Best Data: Consent to Research and the Therapeutic Misconception." *The Hastings Center Report* 17 (2) (1987): 20–24.

Appelbaum, Paul S., Milena Anatchkova, Karen Albert, Laura B. Dunn, and Charles W. Lidz. 2012. "Therapeutic Misconception in Research Subjects: Development and Validation of a Measure." *Clinical Trials* 9 (6) (December): 748–61. https://doi.org/10.1177/1740774512456455.

Binkley, C. E., and Green, B. P. 2021. "Does Intraoperative Artificial Intelligence Decision Support Pose Ethical Issues?" *JAMA Surg* [published online ahead of print, 2021 Jun 16].

Binkley, C. E., J. M. Reynolds, and A. Shuman. 2022. "From the Eyeball Test to the Algorithm - Quality of Life, Disability Status, and Clinical Decision Making in Surgery." *The New England Journal of Medicine* 387 (14): 1325–1328. https://doi.org/10.1056/NEJMms2207408.

Binkley, Charles E., and Bryan Pilkington. 2023. "The Actionless Agent: An Account of Human-CAI Relationships." *The American Journal of Bioethics* 23 (5) (May 4): 25–27. https://doi.org/10.1080/15265161.2023.2191035.

Brummett, Abram, and Lisa Campo-Engelstein. 2021. "Conscientious Objection and LGBTQ Discrimination in the United States." *Journal of Public Health Policy* 42 (2) (June): 322–30. https://doi.org/10.1057/s41271-021-00281-2.

Buchanan, A. 2009a. "Is There a Medical Profession in the House?" In *Justice and Healthcare: Selected Essays*, 175–202. Oxford University Press.

Buchanan, A. 2009b. *Justice and Health Care: Selected Essays*. Oxford: Oxford University Press.

Cavanaugh, T. 2017. *The Hippocrates' Oath and Asclepius' Snake: The Birth of the Medical Profession*. Oxford: Oxford University Press.

CDC. 2024. "Handwashing Facts." Clean Hands, June 13. https://www.cdc.gov/clean-hands/data-research/facts-stats/index.html.

Chi, Michelene T. H., Robert Glaser, and Marshall J. Farr, eds. 2014. *The Nature of Expertise*. 0 ed. Psychology Press. https://doi.org/10.4324/9781315799681.

Christian, F., D. F. Pitt, J. Bond, P. Davison, and A. Gomes. 2008. "Professionalism—Connecting the Past and the Present and a Blueprint for the Canadian Association of General Surgeons." *Canadian Journal of Surgery* 51 (2): 88–91.

Cohen, I. Glenn. 2022. "Travel to Other States for Abortion After *Dobbs*." *The American Journal of Bioethics* 22 (8) (August 3): 42–44. https://doi.org/10.1080/15265161.2022.2089486.

Doernberg, Samuel, and Robert Truog. 2023. "Spheres of Morality: The Ethical Codes of the Medical Profession." *The American Journal of Bioethics* 23 (12) (December 2): 8–22. https://doi.org/10.1080/15265161.2022.2160514.

Elliott, Carl. 2024. *The Occasional Human Sacrifice: Medical Experimentation and the Price of Saying No.* 1st ed. New York, NY: W.W. Norton & Company.

Garcia, Jorge L. A. 2020. "Virtues and Principles in Biomedical Ethics." *The Journal of Medicine and Philosophy: A Forum for Bioethics and Philosophy of Medicine* 45 (4–5) (July 29): 471–503. https://doi.org/10.1093/jmp/jhaa013.

Gwizdala, Crystal. 2022. "When Personal and Professional Morals Clash: Conscientious Objection in Medicine." Accessed July 27, 2024. https://medicine.yale.edu/news-article/when-personal-and-professional-morals-clash-conscientious-objection-in-medicine/.

Harris, P., and J. Ehrnfeld. 2020. "Police Brutality Must Stop." https://www.ama-assn.org/about/leadership/police-brutality-must-stop. Accessed January 2, 2021.

Macintyre, Alasdair. 1991. "Precis of Whose Justice? Which Rationality?" *Philosophy and Phenomenological Research* 51 (1) (March): 149. https://doi.org/10.2307/2107828.

McCradden, Melissa D., James A. Anderson, Elizabeth A. Stephenson, Erik Drysdale, Lauren Erdman, Anna Goldenberg, and Randi Zlotnik Shaul. 2022. "A Research Ethics Framework for the Clinical Translation of Healthcare Machine Learning." *The American Journal of Bioethics* 22 (5) (May 4): 8–22. https://doi.org/10.1080/15265161.2021.2013977.

Meyers, Christopher. 2021. "Deception and the Clinical Ethicist." *The American Journal of Bioethics* 21 (5) (May 4): 4–12.

Miller, Daniel J. 2021. "Justifying Positive Appeals to Conscience: The Debate We Can't Avoid." *The American Journal of Bioethics* 21 (8) (August 3): 79–81. https://doi.org/10.1080/15265161.2021.1940356.

Muyskens, Kathryn Lynn. 2024. "Medical Pluralism as a Matter of Justice." *The Journal of Medical Humanities* 45 (1) (March): 95–111. https://doi.org/10.1007/s10912-023-09809-x.

Nimmon, L., and T. Stenfors-Hayes. 2016. "The "Handling" of Power in the Physician-Patient Encounter: Perceptions from Experienced Physicians." *BMC Medical Education* 16: 114. https://doi.org/10.1186/s12909-016-0634-0.

Pilkington, Bryan. 2020a. "Detention, Dignity, and a Call for Bioethics Advocacy." The Hastings Center's Bioethics Forum, October 22. https://www.thehastingscenter.org/detention-dignity-and-a-call-for-bioethics-advocacy/.

Pilkington, B. 2020b. "Two Kinds of Advocacy," *Professional Formation.* https://webcampus.drexelmed.edu/professionalformation.org/Main.aspx.

Pilkington, Bryan. 2023. "What Is a Physician? Navigating Incommensurable Spheres of Role Morality." *The American Journal of Bioethics* 23 (12) (December 2): 44–46. https://doi.org/10.1080/15265161.2023.2272939.

Pilkington, Bryan, and Ana Campoverde. 2020. "The Bioethics of Translation: Latinos and the Healthcare Challenges of COVID-19." *American Catholic Studies* 131 (3): 11–17. https://doi.org/10.1353/acs.2020.0041.

Pilkington, Bryan, and Charles Binkley. 2022. "Disproof of Concept: Resolving Ethical Dilemmas Using Algorithms." *The American Journal of Bioethics* 22 (7): 81–83. https://doi.org/10.1080/15265161.2022.2087789.

Pilkington, B. C., V. Wilkins, and D. B. Nichols. 2021. "Educating Ethically During COVID-19." *International Journal of Ethics Education.* https://doi.org/10.1007/s40889-021-00120-8.

Pilkington, B., C. Coleman, A. Campoverde, and D. Nichols. 2022. "Vulnerability and Detention in the Time of COVID: An American Failure." *Revue CONFLUENCE: Sciences & Humanités* 1: 91–107. https://www.cairn.info/revue--2022-1-page-91.htm.

Rabinowitz, L. 2018. “Recognizing Blind Spots—A Remedy for Gender Bias in Medicine?” *New England Journal of Medicine* 378: 2253–55. https://doi.org/10.1056/NEJMp1802228.

Savulescu, Julian. 2006. “Conscientious Objection in Medicine.” *BMJ* 332 (7536) (February 4): 294–97. https://doi.org/10.1136/bmj.332.7536.294.

Smith, P. 2019. “Moral Status and Care of Impaired Newborns: An African American Protestant Perspective.” In *Religion and Ethics in the Neonatal Intensive Care Unit*, edited by Ronald M. Green and George A. Little. Oxford: Oxford University Press.

Starr, P. 1982. *The Social Transformation of American Medicine*. New York: Basic Books.

Tsai, D. F. 1999. “Ancient Chinese Medical Ethics and the Four Principles of Biomedical Ethics.” *Journal of Medical Ethics* 25 (4) (August): 315–21. https://doi.org/10.1136/jme.25.4.315.

Ueki, H., Y. Furusawa, K. Iwatsuki-Horimoto, M. Imai, H. Kabata, H. Nishimura, and Y. Kawaoka. 2020. “Effectiveness of Face Masks in Preventing Airborne Transmission of SARS-CoV-2.” *mSphere* 5 (5). https://doi.org/10.1128/msphere.00637.

Vogel, L. 2018. “Should Doctors Talk Politics with Patients?” *Canadian Medical Association Journal* 190 (16): E520–E521. https://doi.org/10.1503/cmaj.109-5589.

Wicclair, Mark R. 2019. “Preventing Conscientious Objection in Medicine from Running Amok: A Defense of Reasonable Accommodation.” *Theoretical Medicine and Bioethics* 40 (6) (December): 539–64. https://doi.org/10.1007/s11017-019-09514-8.

Ziegler, Mary. 2022. “The End of *Roe v. Wade*.” *The American Journal of Bioethics* 22 (8) (August 3): 16–21. https://doi.org/10.1080/15265161.2022.2075962.

Ziegler, Carol, and James Muchira. 2023. “Climate Change: The Ultimate Determinant of Health.” *Primary Care* 50 (4) (2023): 645–55. https://doi.org/10.1016/j.pop.2023.04.010.

MIX
Papier aus verantwortungsvollen Quellen
Paper from responsible sources
FSC® C105338

If you have any concerns about our products,
you can contact us on
ProductSafety@springernature.com

In case Publisher is established outside the EU,
the EU authorized representative is:
Springer Nature Customer Service Center GmbH
Europaplatz 3, 69115 Heidelberg, Germany

Printed by Libri Plureos GmbH
in Hamburg, Germany